Solution Combustion Synthesis of Nanostructured Solid Catalysts for Sustainable Chemistry

Sustainable Chemistry Series

ISSN: 2514-3042

Series Editor: Nicholas Gathergood *(University of Lincoln, UK)*

The concept of Green Chemistry was first introduced in 1998 with the publication of Anastas and Warner's "12 Principles of Green Chemistry". Today, these principles are becoming adopted as general practice in the chemical industries in order to reduce or eliminate the use and generation of hazardous materials, reduce waste, and make use of sustainable resources. New, safer materials and products are being released all the time. Alternative technologies are being developed to improve the efficiency of the chemical industry, while reducing its environmental impact. Sustainable resources are being investigated to replace our reliance on fossil fuels – not only as source of energy but also a source of chemicals — be they feedstock, bulk, or fine. Consideration is now given to the whole life cycle of a product or chemical — from design to disposal. And, as more of the Earth's resources become scarce so new alternatives must be found.

As the world works towards meeting the needs of the present generation without compromising the needs of the future, this series presents comprehensive books from leaders in the field of green and sustainable chemistry. The volumes will offer an excellent source of information for professional researchers in academia and industry, and postgraduate students across the multiple disciplines involved.

Published

Vol. 5 *Solution Combustion Synthesis of Nanostructured Solid Catalysts for Sustainable Chemistry*
edited by Sergio L. González-Cortés

Vol. 4 *Silica-Based Organic–Inorganic Hybrid Nanomaterials: Synthesis, Functionalization and Applications in the Field of Catalysis*
by Rakesh Kumar Sharma

Vol. 3 *Functional Materials from Lignin: Methods and Advances*
edited by Xian Jun Loh, Dan Kai and Zibiao Li

Vol. 2 *Furfural: An Entry Point of Lignocellulose in Biorefineries to Produce Renewable Chemicals, Polymers, and Biofuels*
edited by Manuel López Granados and David Martín Alonso

Vol. 1 *Sorption Enhanced Reaction Processes*
by Alírio Egídio Rodrigues, Luís Miguel Madeira,
Yi-Jiang Wu and Rui Faria

Sustainable
Chemistry
Series

Volume 5

Solution Combustion Synthesis of Nanostructured Solid Catalysts for Sustainable Chemistry

Series Editor

Nicholas Gathergood
University of Lincoln, UK

Sergio L. González–Cortés
University of Oxford, UK

World Scientific

NEW JERSEY · LONDON · SINGAPORE · BEIJING · SHANGHAI · HONG KONG · TAIPEI · CHENNAI · TOKYO

Published by

World Scientific Publishing Europe Ltd.

57 Shelton Street, Covent Garden, London WC2H 9HE

Head office: 5 Toh Tuck Link, Singapore 596224

USA office: 27 Warren Street, Suite 401-402, Hackensack, NJ 07601

Library of Congress Cataloging-in-Publication Data

Names: González-Cortés, Sergio, 1967– editor.

Title: Solution combustion synthesis of nanostructured solid catalysts for sustainable chemistry / Sergio L. González-Cortés, University of Oxford, UK.

Description: London ; Hackensack : World Scientific, 2020. | Series: Sustainable chemistry series, 2514-3042 ; volume 5 | Includes bibliographical references and index.

Identifiers: LCCN 2020026622 | ISBN 9781786348692 (hardcover) | ISBN 9781786348708 (ebook for institutions) | ISBN 9781786348715 (ebook for individuals)

Subjects: LCSH: Catalysts--Synthesis. | Solution combustion synthesis. | Nanostructured materials--Industrial applications. | Green chemistry.

Classification: LCC TP159.C3 S63 2020 | DDC 660.028/6--dc23

LC record available at https://lccn.loc.gov/2020026622

British Library Cataloguing-in-Publication Data

A catalogue record for this book is available from the British Library.

For any available supplementary material, please visit
https://www.worldscientific.com/worldscibooks/10.1142/Q0257#t=suppl

Desk Editors: Britta Ramaraj/Michael Beale/Shi Ying Koe

Typeset by Stallion Press
Email: enquiries@stallionpress.com

Preface

The transition toward greener and more sustainable technology is a direct consequence of the possible devastating effect that anthropogenic emissions and climate change can have over the existence of humankind in our planet. This transition represents one of the grand challenges that our society needs to face in the twenty-first century. To be a sustainable industry, both the natural resources and the residues generated in the process should be treated at rates comparable to the production of supplies and the natural assimilation of residues. Within this context, heterogeneous catalysis and the synthesis of solid catalysts have an enormous positive impact on this transition. The former can enhance the energy efficiency of catalytic processes and decrease the formation of by-products, whereas the latter can make more efficient the catalyst design based on metal abundance and functionality. This would minimize the energy consumption, waste production, and the use of toxic and/or hazardous reagents in the synthesis of catalysts.

Scope of *Solution Combustion Synthesis of Nanostructured Solid Catalysts for Sustainable Chemistry*

This book is a collection of fundamental and applied cutting-edge studies that highlight general and specific principles of the synthesis of nanostructured catalysts through solution combustion synthesis (SCS) and their applications from the perspective of green chemistry. It covers the synthesis of a wide variety of catalytic materials such as nanoparticles of metals, pristine metal oxides, perovskite- and spinel-type metal oxides, solid oxide fuel cell materials, coating of metal oxides, among others using either microwave energy or conventional heating as ignition source. These nanostructured materials are employed as supports, catalyst precursors, or catalysts in heterogeneous catalysis, photocatalysis, and electrocatalysis for the production of hydrogen, biofuels, wastewater treatment, environmental pollution control, and so on.

In *Chapter 1*, several synthesis factors such as metal precursors, fuels, ignition modes, solvents, solution pH, and hard templates, among others have shown strong influences over the greenness of the SCS of advanced catalysts and materials. The below series of criterions were rationally selected to make greener solution combustion (SC)-synthesized catalysts:

1. Low decomposition and reduction temperature of abundant metal nitrates.
2. High solubility of metal precursors and fuel in water or any other environmentally friendly solvent.
3. Neutral acidity of the redox solution.
4. Low tendency of the metal precursor and fuel to contaminate the final catalyst (or material) and minimization of waste quantities of noxious emissions during the combustion process.
5. Utilization of waste-derived precursors and sustainable fuels with high-reducing valence to enhance catalyst performance.
6. Ignition of redox mixture with microwave (or ultrasound) to minimize the energy consumption.
7. Process intensification to minimize the synthesis steps necessary to obtain the final catalyst or material.

Chapter 2 gives a lucid overview of the principles, the applications, and the most significant progress in the microwave-assisted SCS of nanostructured catalysts. The positive aspects of microwave dielectric heating over the SCS method include fast reaction time and high production yield, high purity of nanostructured catalysts, the usual absence of unwanted products, and its sustainable character, in addition to the conventional advantages of microwave technology. All these features find a suitable collocation inside the green chemistry and green engineering fields. It is worth highlighting that most of the catalysts prepared by this method exhibit high surface area and porosity while compared to the other conventional combustion methods. The catalytic activities of the SC-synthesized materials also offered enhanced catalytic performance.

The influence of the SCS of nanostructured materials over different photocatalytic reactions was examined in *Chapter 3*. The authors discussed a vast and wide range of photocatalytic materials such as metal oxides, doped-metal oxides, supported metal oxides synthesized by SC route, and extensively investigated for photocatalytic applications. Primary emphasis in the synthesis of nanostructured photocatalysts has been given over choice of fuel, oxidizer and fuel-to-oxidizer ratio. During the combustion process, these variables control different properties such as the bulk structure, preferential formation of one polymorph over other, surface morphology, crystallinity, particle size, and so on. A promising advantage of the SCS method is engineering the band gap of the semiconducting material to make it more visible light sensitive. Eventually, these properties enhance the photocatalytic efficiency of the materials under visible light exposure.

Chapter 4 describes the advantages, applications, and progress of the SCS of nanomaterials utilized in electrochemistry. The SCS method is attractive for electrochemistry applications, owing to the high specific surface area, the nanocrystalline nature of the obtained products, and its capability to synthesize metastable compounds. A variety of materials derived from the SCS has been widely used in oxygen reduction reaction (ORR), oxygen evolution reaction (OER), hydrogen evolution reaction (HER), photoelectrochemistry (PEC), electrocatalytic ethanol oxidation, and supercapacitors. Remarkable progresses have been made in all these research areas; however, the mechanism pathways of the SCS are still needed to be understood, which is the prerequisite of fine tuning the phase compositions and nanostructures of the corresponding electrochemical catalysts/electrodes. Furthermore, the synthesis of high-performance electrocatalysts in thin films or arrays through the SCS method also needs to be enhanced.

A comprehensive review to solid oxide fuel cells, the SC method, and an overview of the various solid oxide fuel cell (SOFC) materials and coatings prepared by SC method is given in *Chapter 5*. Various synthesis routes such as solid-state method, coprecipitation, sol-gel, the SC method, and so on, are used for the synthesis of SOFC materials. Among these methods, the SC process is the most popular approach that is employed for the synthesis of an array of oxide materials for SOFC application because of its simplicity and versatility. The SC method has been used for the synthesis of cathode, anode, electrolyte, and interconnect materials using different fuels, the mixture of fuels, and varying oxidizer:fuel ratios. Experimental evidences have showed that by varying the SC parameters and fuels, it is possible to synthesize oxide powders possessing nano and micron size that are suitable for fabrication processes such as tape casting and plasma spraying, respectively.

Chapter 6 overviews the current progress of SC-synthesized catalysts in the production of clean energy and environmental pollution control for the sustainable development of advanced catalytic processes and gives an overall conclusion and path forward of the SCS approach. A large variety of catalysts such as noble metals, base metals — particularly nickel — and bimetallic formulations mainly supported on ceria and alumina have been investigated in the production of hydrogen from a variety of hydrocarbons. These catalysts are mainly obtained from the SC-synthesized mixed-metal oxides as catalyst precursors, which after reduction produce nanosized metal particles and strong metal-support interactions that enhance the activity, hydrogen selectivity, and catalytic stability. It is also an extended practice to make the catalyst support by the SC method and then depositing the metal precursor(s) by impregnation. Most of the catalyst formulations have been examined in the reforming of hydrocarbons, water–gas shift reaction, and even cracking (or catalytic decomposition) of methane under nonoxidative atmosphere. The SC-synthesized nanopowder catalysts or structured catalysts have also showed advantages for bio-hydrogen production, methane oxidation, soot combustion, hydrodesulfurization, and NO_x abatement among many others applications.

It has been a real privilege to work with some of the most recognized scientists and academics involved in the field of the SC method applied to catalysis. I would like to express my sincere thanks to the lead authors and the coauthors for their commitment to deliver such high-value contributions and their support to produce a high-quality book. I also gratefully acknowledge the available facilities of the Inorganic Chemistry Laboratory at the University of Oxford during the development of this project. I would also like to thank the reviewers for their high standard and very valuable input to enhance the presentation of this book. Finally, I very much appreciate the opportunity that World Scientific and Professor Nicholas Gathergood gave me to edit this material in the *Sustainable Chemistry Series*. I hope that the students, researchers, and general readership find this book a valuable reference for green chemistry, nanochemistry, materials science, and chemical engineering.

Sergio L. González-Cortés
Oxford, October 15, 2019

About the Editor

Sergio L. González-Cortés received his Licentiate in Chemistry, Cumlaude honour, at the University of Los Andes (ULA, Venezuela), in 1993. After 8 years working as assistant-associate professor at ULA and conducting research independently in methane valorization, he moved to the United Kingdom, where he received his D. Phil in Heterogeneous Catalysis from the University of Oxford in 2005. Sergio was awarded a Scholarship from the University of Los Andes (ULA, Venezuela) and the National Funds of Science, Technology and Innovation (Venezuela) to complete his PhD at Oxford University. He was a visiting scholar at the University of California at Riverside and at the University of Oklahoma (USA). He was also an invited scholar to attend Material Characterization Course in Brazil. Sergio has published more than 60 papers, several reviews, book chapters, and patents that are currently in the development stage.

He joined Oxford Catalysts Company (now Velocys) in 2007 as Senior Research Scientist to lead the development of hydrotreating catalysts for ultraclean diesel fuel and then Carbon Trust (Future Blends Ltd) in 2012 as Principal Research Scientist to conduct research in hydroprocessing of bio-oils to biofuels. He currently works at the University of Oxford as Senior Departmental Research Officer, focusing on the potential applications of dielectric heating in a variety of catalytic processes of interest in the petrochemical industry. Sergio's research interests can be divided into three specific areas:

1. Integration of new and advanced catalysts in conventional processes or new catalytic reactions for applications in renewable energy production.
2. Development of innovative approaches to synthesize advanced solid (nano) catalysts for upgrading unconventional fossil resources.
3. Fundamental (in situ or ex situ) studies focused on the influence of the synthesis approach over the physical and chemical properties of the solid catalysts and their relationships with the catalytic performances.

Chapter Contributor Information

Chapter 1

Sergio L. González-Cortés (Editor and Corresponding Author)

Email: sergio.gonzalez-cortes@chem.ox.ac.uk; slgoncor@gmail.com

Postal Address: Inorganic Chemistry Laboratory, Department of Chemistry, University of Oxford, South Parks Road, Oxford, OX1 3QR, United Kingdom

Afrah M. Aldawsari

Email: afrah.aldawsari@gmail.com; amaldossari@kacst.edu.sa

Postal Address: Umm Al-Qura University, P.O. Box 715, Makkah 21955, Kingdom of Saudi Arabia

Serbia Rodulfo-Baechler

Email: mariserbrod@gmail.com

Postal Address: Inorganic Chemistry Laboratory, Department of Chemistry, University of Oxford, South Parks Road, Oxford, OX1 3QR, United Kingdom

Chapter 2

Deshetti Jampaiah

Email: sampathdeshetti@gmail.com

Postal Address: Catalysis and Fine Chemicals Department, CSIR-Indian Institute of Chemical Technology, Uppal Road, Hyderabad 500007, India

Perala Venkataswamy

Email: pvschemou07@gmail.com

Postal Address: Catalysis and Fine Chemicals Department, CSIR-Indian Institute of Chemical Technology, Uppal Road, Hyderabad 500007, India

Benjaram M. Reddy (Corresponding Author)

Email: bmreddy@iict.res.in

Postal Address: Catalysis and Fine Chemicals Department, CSIR-Indian Institute of Chemical Technology, Uppal Road, Hyderabad 500007, India

Chapter 3

Sounak Roy (Corresponding Author)

Email: sounak.roy@hyderabad.bits-pilani.ac.in

Postal Address: Department of Chemistry, BITS Pilani, Hyderabad Campus, Jawahar Nagar, Shameerpet Mandal, Hyderabad 500078, India

Swapna Challagulla

Email: challagulla.swapna@gmail.com

Postal Address: Department of Chemistry, BITS Pilani, Hyderabad Campus, Jawahar Nagar, Shameerpet Mandal, Hyderabad 500078, India

Chapter 4

Wei Wen (Corresponding Author)

Email: wwen@hainu.edu.cn

Postal Address: College of Mechanical and Electrical Engineering, Hainan University, Haikou 570228, P. R. China

Jin-Ming Wu

Email: msewjm@zju.edu.cn

Postal Address: State Key Laboratory of Silicon Materials and School of Materials Science and Engineering, Zhejiang University, Hangzhou 310027, P. R. China

Chapter 5

S.T. Aruna (Corresponding Author)

Email: aruna_reddy@nal.res.in

Postal Address: Surface Engineering Division, Council of Scientific and Industrial Research-National Aerospace Laboratories, Bangalore 560017, India

S. Senthil Kumar

Email: ssenthil@nal.res.in; sssenkum@gmail.com

Postal Address: Surface Engineering Division, Council of Scientific and Industrial Research-National Aerospace Laboratories, Bangalore 560017, India

Chapter 6

Sergio L. González-Cortés (Editor and Corresponding Author)

Email: sergio.gonzalez-cortes@chem.ox.ac.uk; slgoncor@gmail.com

Postal Address: Inorganic Chemistry Laboratory, Department of Chemistry, University of Oxford, South Parks Road, Oxford, OX1 3QR, United Kingdom

Contents

https://doi.org/10.1142/9781786348708_0001

Chapter 1

Green Chemistry for Solution Combustion Synthesis of Advanced Catalysts and Materials

Sergio L. González-Cortés[*,‡], Afrah M. Aldawsari[†,§],
and Serbia Rodulfo-Baechler[*,¶]

[*]*Inorganic Chemistry Laboratory, Department of Chemistry,
University of Oxford, South Parks Road,
Oxford, OX1 3QR, United Kingdom*
[†]*Umm Al-Qura University, P.O. Box 715, Makkah 21955,
Kingdom of Saudi Arabia*
[‡]*sergio.gonzalez-cortes@chem.ox.ac.uk; slgoncor@gmail.com*
[§]*afrah.aldawsari@gmail.com; amaldossari@kacst.edu.sa*
[¶]*mariserbrod@gmail.com*

1.1. Introduction

Nowadays, eco-friendly approaches to produce nanomaterials, particularly nanocatalysts, have been evolving to what is now known as *Green Chemistry*. *Green Chemistry* is embodied in 12 principles, which can be briefly described as:

1. Prevention of waste
2. Atom efficiency
3. Less hazardous/toxic chemicals
4. Safer products by design
5. Innocuous solvents and auxiliaries
6. Energy efficient by design
7. Preferably renewable raw materials
8. Shorter syntheses (avoid derivatization)
9. Catalytic rather than stoichiometric reagents

10. Design products for degradation
11. Real-time monitoring and process control
12. Inherently safer chemistry (Sheldon *et al.*, 2007)

These 12 principles guide a general trend within the chemistry community and beyond. We are required to substitute traditional raw materials by renewable and waste-derived matters, but also to use sustainable energy resources to mitigate the generation of hazardous and toxic waste materials and also to enhance the energy efficiency of the process.

The recent progresses for producing energy from sustainable resources (biogases/hydrogen, biofuels, solar energy, etc.) are mainly based on processes that involve at least a catalytic stage (e.g., heterogeneous catalysis, homogeneous catalysis, electro-catalysis, photocatalysis, and biocatalysis). These advances have been achieved through innovative synthesis methods, computational chemistry, and operando spectroscopy that produce advanced solid catalysts by design rather than trial and error. Within the context of solid catalysts for commercial applications, they need to be an active, selective, stable, and cost-effective material for a determined catalytic process (Rodulfo-Baechler *et al.*, 2018; González-Cortés *et al.*, 2014; Munnik *et al.*, 2015). The best environmentally friendly catalyst synthesis method must be able to produce a catalytic material with appropriate textural properties (i.e., sufficiently high surface area and uniform pore distribution), suitable mechanical strength, high attrition, efficient regeneration (rejuvenation), and recovery from the reaction medium. Furthermore, the solid catalysts should be based on base metals, rather than noble metals (Ludwig and Schindler, 2017), to enhance their greenness and sustainability properties.

The preparation methods of solid catalysts can be grouped into two major groups as shown in Fig. 1.1, in which is also illustrated several types of catalysts. Methods able to produce bulk (unsupported) catalysts are based on precipitation, decomposition, combustion reactions, among others. Methods that introduce and fix the active phase or catalyst functionality onto a preexisting solid (carrier), or coated catalysts, through a process intrinsically dependent on

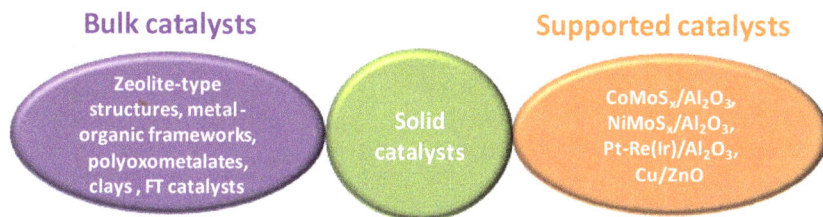

Figure 1.1. Two principal categories of solid catalysts prepared through a variety of methods.

the surface area of the support (impregnation, grafting/anchoring methods) and dimension of the cavity (confined/"ship-in-a-bottle" method) (Deutschmann *et al.*, 2009; Munnik *et al.*, 2015). The best synthesis method must be able to:

1. Produce a catalytic material with suitable physical and chemical properties.
2. Operate for relatively long periods with high activity and selectivity to the targeted product.
3. Be facilely recovered or regenerated.

Other synthesis methods such as sol-gel synthesis (Heinrichs *et al.*, 2007), deposition–precipitation (de Jong, 2009), flame hydrolysis, precipitation, among others (Hagen, 2015), and even solution combustion synthesis (SCS) (González-Cortés and Imbert, 2013) are also classified into these categories (i.e., methods that produce bulk catalysts and methods that introduce and fix the active phase onto a preexisting solid).

The self-sustained reaction synthesis (SRS), or combustion synthesis (CS), is a versatile approach to prepare a plethora of nanomaterials/nanocatalysts such as single- and mixed-metal oxides, ceramic, metals, alloys, sulphides, among others. This method takes advantage of an exothermic and usually very rapid and self-sustained chemical reaction. Its key feature is the fact that the heat required to drive the reaction is provided by the reaction itself and not by an external source, nevertheless, it is necessary to reach an ignition temperature to start the reaction. Two main methods can be distinguished: (1) self-propagating high-temperature synthesis (SHS)

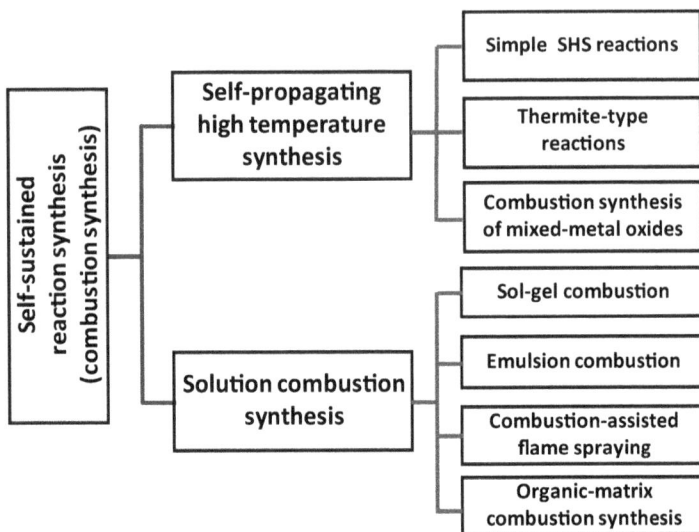

Figure 1.2. Classification of self-sustained reaction syntheses for preparing advanced nanomaterials. The ignition temperature can be achieved by either conventional or unconventional heating, whereas the combustion initiation mode can be through volume combustion or self-propagating modes.

and (2) SCS (Fig. 1.2). These routes differ mainly on the reactivity of the gas phase and the physical state of the reactants and products. The SHS method was initially proposed and developed by Russian scientists in 1967 as an alternative process for producing advanced materials (Merzhanov, 1996, 2004). The mixed reactant powders are ignited either in the propagating mode (i.e., ignited locally at one point of the sample) or simultaneous mode (by heating the whole sample to the ignition temperature). The heat generated can raise the temperature of the reactant powder by hundreds of degrees, facilitating not only the product formation but also its sintering within a relatively short reaction time (Moore and Feng, 1995; Morsi, 2012). According to Moore and Feng (1995), the SHS method can be classified as simples SHS reaction, thermite-type reactions, and CS of mixed-metal oxides, see Fig. 1.2. Furthermore, a large variety of SHS-related hybrid processes have been developed (Morsi, 2012).

The SCS approach was proposed by Kingsley and Patil (1988). It consists of using a saturated aqueous solution of the desired metal salts (nitrates are generally preferred because of their oxidizing property and high solubility in water) and a suitable organic fuel as reducing agent (e.g., urea, citric acid, and glycine). The redox mixture is ignited and eventually combusted in a self-sustained and fast combustion reaction, resulting, usually, in nanocrystalline materials (Bera and Aruna, 2018; Patil *et al.*, 2008; Specchia *et al.*, 2010). A variety of derived approaches such as sol-gel combustion (Butkute *et al.*, 2018; Sutka and Mezinskis, 2012), emulsion CS (Chandradass *et al.*, 2010), combustion-assisted flame spraying or spray (solution) CS (Trusov *et al.*, 2016; Xanthopoulou *et al.*, 2017; Yu *et al.*, 2015), and organic-matrix CS (González-Cortés *et al.*, 2004; González-Cortés *et al.*, 2006) have been proposed. Colloidal SCS (Voskanyan *et al.*, 2016) and other methods (Mukasyan and Dinka, 2007) can be included into the previous categories. Some of these approaches are suitable for particular applications that include solid catalysts, supercapacitors, semiconductors, optical material, energy storage, and so on (Li *et al.*, 2015; Varma *et al.*, 2016).

A relatively large number of literature reviews focused on the fundamentals, properties, progress, and advantages (and disadvantages) of the SCS approaches for synthesizing advanced nanomaterials have been published (Aruna and Mukasyan, 2008; Bera and Aruna, 2018; Carlos *et al.*, 2020; Deganello and Tyagi, 2018; Frikha *et al.*, 2019; González-Cortés and Imbert, 2013; Mukasyan and Dinka, 2007; Mukasyan *et al.*, 2015; Patil *et al.*, 2008; Thoda *et al.*, 2018; Varma *et al.*, 2016), reflecting the impact that this relatively simple methodology has had on different fields. However, very little progress has been made with regard to its eco-friendly properties and potential utilization as a sustainable and green approach. The aim of this chapter is to review the current methods and to identify the potential developments of the SCS from a point of view of sustainability and "greenness," to mitigate the generation of hazardous and toxic waste materials and to enhance the energy efficiency of the combustion process. This chapter looks not only at the challenge of producing advanced and eco-friendly nanocatalysts and general nanomaterials through the SCS, but also from the possible technologic barriers of future developments framed within the context of a circular economy.

1.2. Characteristics of Solution Combustion Synthesis

The SCS method is usually carried out over an aqueous solution of a metal nitrate as oxidizing agent and a suitable organic fuel employed as reducing agent. The ignition of the redox mixture at a determined temperature, below 500°C, initiates a self-propagating exothermic reaction that sustains high temperatures for a sufficient period of time to decompose all the organic material and metal salts. The final product is usually a nanomaterial with a large specific surface area as a consequence of the large amount of gases produced during the combustion reaction (González-Cortés and Imbert, 2013). To illustrate this process, below is given the stoichiometric redox reaction of nickel nitrate as the oxidizer and urea as the fuel to produce nickel oxide, Eq. (1.1).

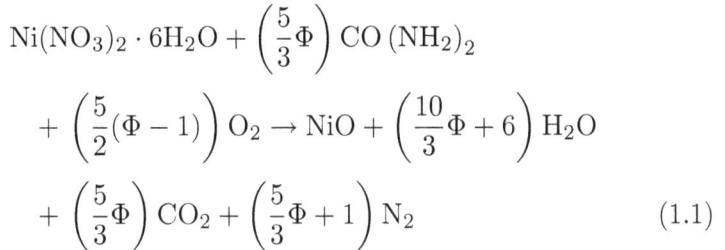

$$Ni(NO_3)_2 \cdot 6H_2O + \left(\frac{5}{3}\Phi\right) CO\,(NH_2)_2$$

$$+ \left(\frac{5}{2}(\Phi - 1)\right) O_2 \rightarrow NiO + \left(\frac{10}{3}\Phi + 6\right) H_2O$$

$$+ \left(\frac{5}{3}\Phi\right) CO_2 + \left(\frac{5}{3}\Phi + 1\right) N_2 \tag{1.1}$$

Φ is related to the equivalence ratio of the reactants (Jain *et al.*, 1981) and it is directly proportional to the fuel-to-oxidizer ratio. The optimal stoichiometric composition (or equivalence ratio) of the redox mixture is obtained when no molecular oxygen is required or the reducing and oxidizing species are equivalent (i.e., $\Phi e = 1$). When Φ is below unity ($\Phi < 1$), the redox mixture is under fuel-lean regime and molecular oxygen is produced. On the other hand, when $\Phi > 1$, the redox mixture is under fuel-rich condition, requiring molecular oxygen to fully convert the fuel.

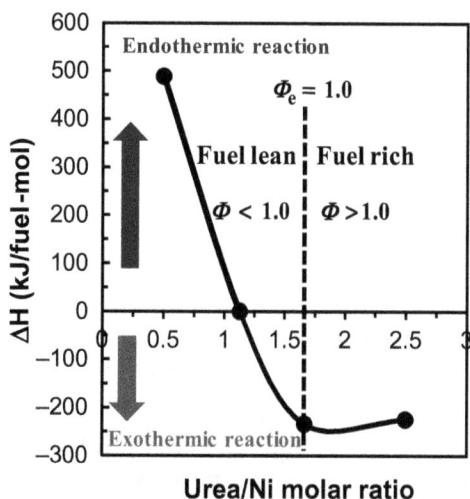

Figure 1.3. Standard reaction enthalpy versus fuel-to-oxidizer ratio for nickel nitrate hydrate–urea redox mixture. Adapted from González-Cortés and Imbert (2013).

The dependence of the fuel-to-oxidizer ratio over the reaction enthalpy (ΔH) is shown in Fig. 1.3. At fuel-lean condition, the reaction is endothermic and becomes exothermic at urea/Ni molar ratios above 1. At optimal stoichiometric composition ($\Phi e = 1$) and fuel-rich condition, the combustion reaction releases the maximum energy. Note that the solution combustion process produces not only the nanostructured material, but also steam, nitrogen/nitrogen oxides, carbon monoxide and carbon dioxide. Therefore, the advantage of a self-propagating reaction, reflected in shorter reaction time and possibly high energy efficiency, is partially eclipsed because of the CO_2 generation, whose concentration is directly proportional to the amount of fuel used.

An important feature of the SCS approach is the extremely high temperatures that can be achieved within very short periods. It is therefore reasonable to assume that a thermally isolated system exists because there is very little time for the heat to dissipate to its surroundings. Therefore, the maximum temperature achieved by the reaction products is assumed to be adiabatic temperature

(T_{ad}), which can be calculated using evolutions of heat capacity with temperature (Eq. (1.2)):

$$-\Delta H_r^\circ = \int_{T_o}^{T_{ad}} \left(\sum n \cdot C_{p(\text{products})}(T) \right) \cdot dT, \qquad (1.2)$$

where ΔH_r° is the enthalpy of reaction at T_o (298 K), n is the stoichiometric coefficient of the products, and C_p is the heat capacity of the products at constant pressure.

Considering the enthalpy change involved in the chemical reaction for NiO synthesis at optimal stoichiometric composition ($\Phi e = 1$), the stoichiometric coefficients, the heat capacity, and in general the relevant thermodynamic data (Colomer *et al.*, 1999), an adiabatic temperature of ~973 K (700°C) was obtained. This is significantly higher than the temperature required for decomposing nickel nitrate to nickel oxide (Brockner *et al.*, 2007). Patil *et al.* (2008) reported a combustion temperature of 1773 K (1500°C) for the synthesis of α-Al$_2$O$_3$ by the SCS in agreement with the adiabatic temperature (i.e., 1700 K; 1427°C).

It is worth remarking that through a temperature–time (T–t) profile of the redox mixture, it is possible to find out the influence of the loading and nature of the metal oxide over the combustion process as illustrated in Fig. 1.4. The temperature profile for 5 wt% cobalt oxide loading over alumina shows a markedly lower temperature (244°C) than the catalyst with high Co content (295°C), whereas the ignition temperature did not change markedly (i.e., 205°C–215°C). The alumina-supported 10 wt% NiO catalyst, on the other hand, shows a significant higher ignition temperature (~275°C) than the equivalent alumina-supported cobalt oxide catalyst, whereas the maximum temperatures were fairly comparable for both samples. When the difference between the maximum temperature and the ignition temperature (ΔT) is considered, it becomes evident that the combustion process, particularly the exothermicity, for the redox mixture of Co precursor-urea (90°C) is markedly different than that for Ni precursor-urea mixture (37°C) as a consequence of their different decomposition temperatures and intermediate products

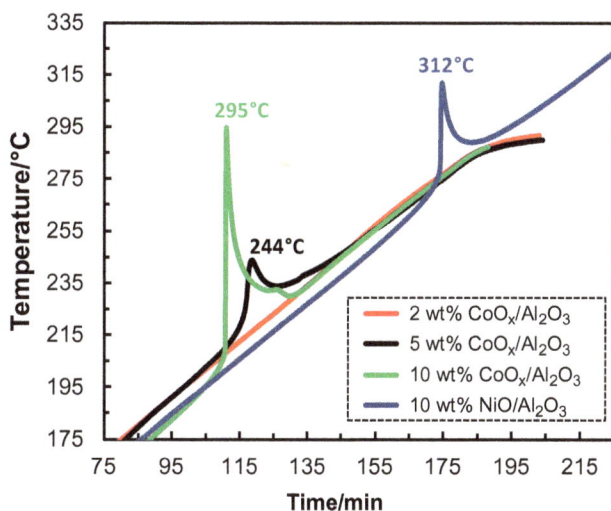

Figure 1.4. Temperature–time (T–t) profiles for alumina-supported Ni and Co oxides. Samples synthesized with urea/Ni molar ratio of 2, combusted under static air condition and heating rate of $1°C\ min^{-1}$.

(Manukyan *et al.*, 2013). It is also expected that the generation of high oxidation states, that is, Co_3O_4 instead of cobalt (II) oxide, alongside the fuel and the presence of γ-alumina would affect the thermolysis of the corresponding nitrates.

Another important aspect of the solution combustion process is the nature of the fuel and the fuel-to-oxidizer ratio to assist the synthesis of nanostructured catalysts. The mechanisms can vary from flaming (gas phase) to nonflaming (smouldering and heterogeneous) types (Patil *et al.*, 1997; Varma *et al.*, 2016), where gaseous products such as nitrogen oxides, HNCO, NH_3, CO_2, and so on, can take place in the flaming pathway, whereas the condensed and gas-phase reactions can simultaneously occur in the smouldering mechanism. The role of the fuel in the SCS is quite versatile since it can act not only as combustion assisting-agent (i.e., reducing agent), but also as chelating and microstructure-controlling agents (Fig. 1.5), depending on its structural characteristics, concentration, acidity (or basicity), and type of metal precursor.

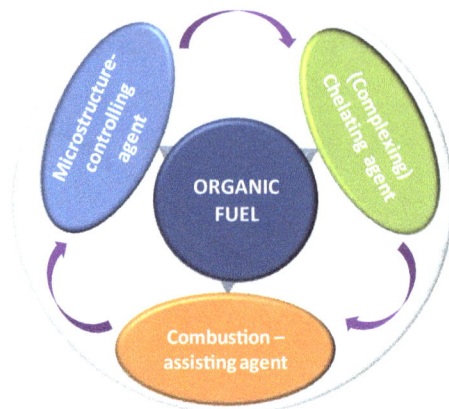

Figure 1.5.   Principal roles of the organic fuel in the SCS.

The organic fuel acting as microstructure-controlling agent can either produce a gel-like network in solution and a final material with hierarchical microstructure after the combustion process (G. Xanthopoulou *et al.*, 2018) or replicate the structure of a hard template to produce an original template-like microstructure (Voskanyan *et al.*, 2016). An organic fuel can also produce a viscous solution containing a metal nitrate mixed with a complexing or chelating agent, this redox mixture undergoes self-propagating combustion between 200°C and 300°C with citric acid as fuel (González-Cortés *et al.*, 2015; Mali and Ataie, 2005).

The organic fuel for the SCS is usually soluble in water; however, organic solvent can also be used to improve fuel solubility. The fuel in aqueous solution can also act as metal precursor dispersing through hydrogen bond interaction, and it can also act eventually as combustion-assisting agent on the combustion process. The combination of fuel with metal precursor produces redox mixture that ignites a temperature below 500°C to produce a self-propagating combustion reaction. Figure 1.6 shows the T–t profile for a metal complex network produced on the mixture of an organic fuel with a metal nitrate treated at mild temperature (below 120°C). As the temperature of the redox mixture is increased, the gel network

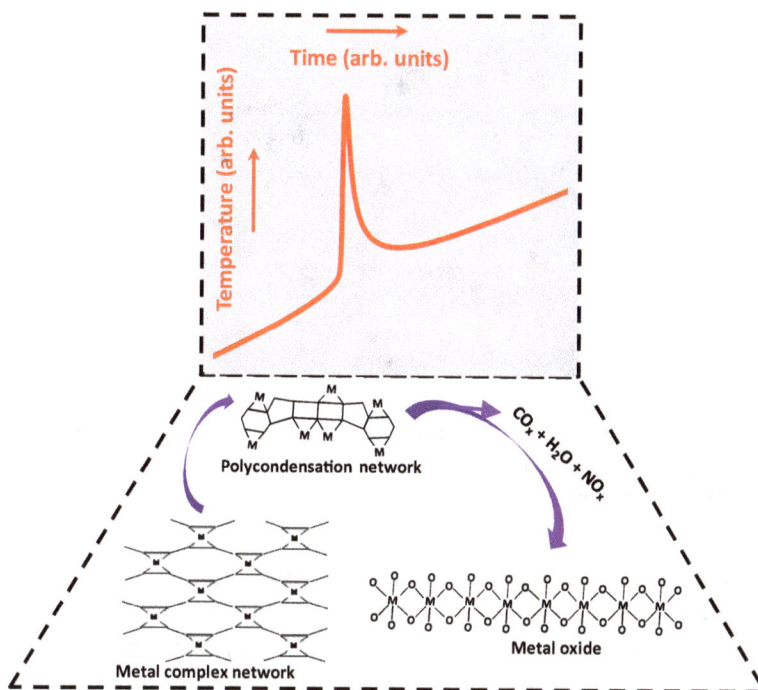

Figure 1.6. Different steps involved in the SCS of metal oxides.

partially collapses toward a polycondensation network that alongside the counterion (i.e., nitrate) achieves the ignition temperature and hence a self-sustained combustion reaction.

The presence of a polycondensation matrix can enhance the nucleation stage with well-dispersed numerous sites that ensure the formation of small crystallite size (Danks *et al.*, 2016). The emission of large amounts of gases upon the combustion also contributes to the textural properties of the nanostructured metal oxide.

1.3. Metrics for Sustainable Development

According to the Brundtland report (1987), a sustainable development enables the present generation to meet its own needs, without compromising the ability of future generations to meet their needs. Sustainable development is usually associated with

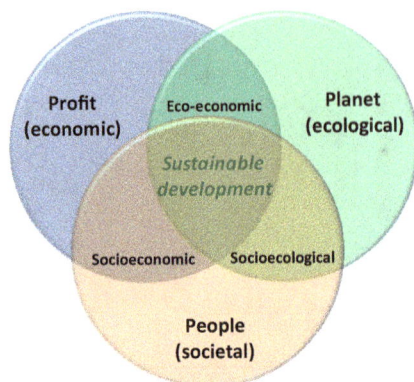

Figure 1.7. Sustainable development model based on three goals: people (societal), planet (ecological), and profit (economic). Adapted from Azapagic and Perdan (2000) and Sheldon (2018).

three components: people (societal), planet (ecological), and profit (economic) as illustrated in Fig. 1.7. The overlap of two components or indicators produces two-dimensional metrics (i.e., eco-economic, socio-ecological, and socioeconomic), whereas the interception of three circles corresponds to optimal sustainable development.

The implementation of a sustainable development approach requires that natural resources should be used at rates that can be replaced naturally over the long term and the generation of wastes should not be faster than the rate of their remediation (Graedel, 2002). For instance, the utilization of nonrenewable resources such as methane, crude oil, and coal to produce energy or chemicals is not sustainable in the long term. These resources are consumed at a much higher rate than they are naturally recovered and the high rates of carbon dioxide generation are markedly superior to the rate at which nature can assimilate it (Sheldon, 2016).

1.3.1. *Greenness Metrics*

As scientists and engineers usually include economic, technical, and social metrics into new process design, they should also incorporate sustainability metrics, a fairly standard practice in the pharmaceutical industry (Jimenez-Gonzalez *et al.*, 2012; McElroy *et al.*, 2015).

The sustainability of a chemical product, chemical process, or activity throughout its life cycle is strongly dependent on the environment, characteristics of the technology, and even economic factors. In fact, the measurement of the greenness of a chemical process needs a holistic approach, in which not only the quantity of produced waste but also metrics associated with process economy, materials, resources, energy, carbon footprint, and so on, are considered. An integrated method that takes into account all these factors, including production/extraction/purification of raw materials, transportation, distribution, utilization, and final application, is known as Life Cycle Assessment (LCA) (Jimenez-Gonzalez and Overcash, 2014; Roy *et al.*, 2009). This approach presents four interrelated stages: objective and scope definition, inventory analysis, impact assessment, and data interpretation. The goal and the factors considered of an inventoried and assessed process within the entire supply chain (cradle-to-grave assessment) or single process (gate-to-gate assessment) will define the reliability of LCA. This sort of analysis can be complemented with other tools (e.g., risk assessment, site-specific environmental assessment, cost assessment, and others) to provide better integration into the decision-making process (Curran, 2013). LCA output is highly sensitive to the quality and quantity of data available; hence, the definition of dataset boundaries will largely affect the outcome of an LCA (Shirvani *et al.*, 2011). In fact, the collection of data for an LCA is not a simple task considering the large amount of data required from a variety of sources, which are not always collected for LCA purposes.

The large amount of work and cost for the LCA methodology can make it more relevant to products and processes at high technology readiness stages and at the commercial application level. Hence, the greenness of research in the initial stage of development can be assessed with simple mass-based green metrics such as atom economy (AE), environmental (E) factor, process mass intensity (PMI), among others (Calvo-Flores, 2009; Curzons *et al.*, 2001; Sheldon, 2018). A selected group of useful green metrics is listed in Table 1.1. The AE introduced by Trost (1991) considers the stoichiometric quantities of the starting materials and the final product(s) of a

Table 1.1. Selected mass-based metrics to measure greenness of chemical reactions or processes.

Metric	Formula	Metric impact
AE	$AE = \dfrac{MW\,(product) \times 100}{\sum MW\,(reactants)}$	The larger AE value is associated with higher content of the reactants in the final product (Trost, 1991)
EF	$EF = \dfrac{\sum Mass\,of\,wastes}{Mass\,of\,product}$	The environmental impact of the process is minimized at lower EF or lower production of wastes (Sheldon, 1992)
PMI	$PMI = \dfrac{Total\,mass\,in\,process}{Mass\,of\,product}$	The closer the PMI to the unity, the lower the environmental effect of a particular process (Curzons *et al.*, 2001)
CE	$CE = \dfrac{Mass\,of\,carbon\,in\,product \times 100}{Total\,mass\,of\,carbon\,in\,reactants}$	The larger the CE of a process, the higher the content of carbon in the final product (Constable *et al.*, 2002)

chemical reaction. It does not take into account solvents and auxiliary chemicals that can play an important role in the synthesis yield of the final product. AE is a very straightforward metric particularly relevant for developing new synthetic pathways, as the previous step of experimentation; that maximize the desired product and minimize the formation of secondary products (Li and Trost, 2008).

The E factor is a second earlier metric proposed by Roger Sheldon that considers the quantity of waste produced for a determined mass of the targeted product (Sheldon, 1992). Water-derived waste was excluded in the original definition of EF but the current trend in the chemical industry is to include it to lead greenness innovation that results in a reduction of overall waste (Sheldon, 2017). Another important metric is the mass intensity (MI) introduced by Curzons *et al.* (2001) and renamed as PMI by The Green Chemistry Institute Pharmaceutical Round Table. PMI describes the ratio of the total mass of materials (i.e., reactants, reagents, solvents used for reaction and purification, and catalysts) used to produce a certain mass of the product. This metric is used to benchmark the environmental footprint of pharmaceutical processes and to assess the greenness of a process for APIs (Jimenez-Gonzalez *et al.*, 2011). Carbon efficiency (CE) is another useful metric that quantifies the wt% of carbon in the reactants that remain in the final product (Constable *et al.*, 2002). This metric is particularly relevant for the SCS process, because it can measure the carbon footprint for catalyst preparation methods discussed in Section 1.4.

1.4. Sustainability of SCS

Green synthesis methodology of heterogeneous catalysts should not only minimize their environmental impact by reducing waste and energy consumption, but also should use safer and benign reactants and auxiliary chemicals. This also needs to produce advanced and commercially viable solid catalysts for process intensification that protect the environment and boost societal benefits. These fundamental concepts framed within the 12 principles of green chemistry (Anastas and Kirchhoff, 2002; Anastas *et al.*, 2001) can be considered as the vertebral column of *green and sustainable catalysis* (GSC), which is centered on the overall sustainability of catalyst synthesis/recovery/regeneration and catalyst application. This holistic approach takes into account the development of sustainable technologies not only for obtaining the end products (i.e., chemicals, fuels, and polymers), but also for synthesizing and regenerating solid catalysts, following all these stages by *in-situ* analysis/characterization and real-time process control. The economic, societal, and ecological aspects are the main core of this sustainable development.

There are a variety of methodologies to synthesize solid catalysts, but there has not been a proper assessment to find out the sustainability of these approaches despite a large volume of catalysts being annually produced/consumed worldwide (Bravo-Suarez *et al.*, 2013; de Jong, 2009). Among these processes, the SCS of heterogeneous catalysts is a relatively new approach whose main components: metal precursor (particularly metal nitrate), organic fuel, solvent alongside ignition source, catalyst carrier, and, in general, the overall process all need to be considered in an assessment of sustainability.

1.4.1. *Nitrate Salts as Metal Precursors*

Metal nitrate as the metal precursor and as an oxidizing agent is a well-known reagent for synthesizing solid catalysts via the SCS method (González-Cortés and Imbert, 2013). Other metal precursors such as hydroxides (Kaur *et al.*, 2016), alkoxides (Burgos-Montes *et al.*, 2006; Muresan *et al.*, 2015), acetates (Nair *et al.*, 2009; Pourgolmohammad *et al.*, 2017b), oxalates (Sekar *et al.*, 1992; Deshpande *et al.*, 2004), and even acetylacetonates (Manjunath *et al.*, 2018; Ortega-San Martín *et al.*, 2016) have also been used, but they have a reducing character instead of oxidizing property and an additional oxidant is required to obtain a suitable redox mixture.

The metal nitrates can be classified into ionic and covalent compounds based on the polarization of the electronic cloud of the nitrate ion by the high charge density of the metal ion and through the back-donation of the electron as shown in Fig. 1.8. Ionic metal nitrates produce hydrated compounds, which have high solubility in polar solvents, particularly water, whereas anhydrous base metal nitrates and their eutectic mixtures are suitable for study as molten salts (Jones, 1973).

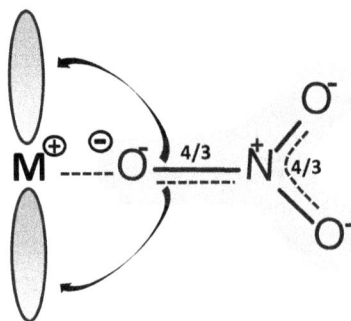

Figure 1.8.   Back-donation of the π-electronic cloud in nitrate ions to the unfilled *d*-orbitals in transition metals. Adapted from Yuvaraj *et al.* (2003).

Metal nitrates as precursors of metal and metal oxide catalysts are widely used in catalyst synthesis because they have high solubility in water and low tendency to contaminate the final catalyst compared to other inorganic salts such as sulphates, phosphates, and halides, and besides they are cheaper. Furthermore, metal nitrates, particularly transition metals, can be decomposed to metal oxides or reduced to metals under relatively mild temperatures as shown in Fig. 1.9. In fact, those salts can be considered as *sustainable metal precursors* because of their relatively low decomposition and reduction temperatures (i.e., high energy efficiency) (Yuvaraj *et al.*, 2003), their minimum safety concerns, and relatively high abundance, particularly for the first-row transition metals (Kaushik and Moores, 2017; Ludwig and Schindler, 2017).

However, on decomposition, they produce substantially high concentrations of waste NOx which limits their greenness. The utilization of gas capture or NO$_x$ abatement can overcome the current limitation of this type of metal precursors. It is worth remarking that the utilization of waste or natural-derived metal precursors such as battery-extracted Zn–Mn precursors (Gabal *et al.*, 2016), rusted

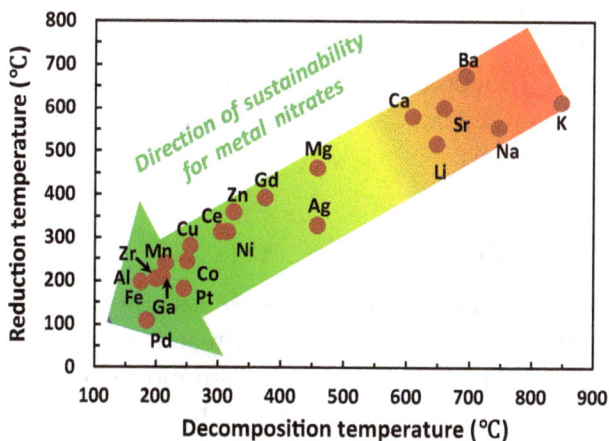

Figure 1.9. Sustainability trend of alkali, alkaline, and transition metal nitrates according to the direct relationship between the decomposition temperatures in H$_2$ (i.e., reduction temperature) and air (i.e., decomposition temperature), based on the data reported elsewhere (Yuvaraj *et al.*, 2003).

iron-containing materials (Deganello *et al.*, 2019), and even chicken eggshell-extracted Ca precursor (Choudhary *et al.*, 2015) have been used for the sustainable synthesis of mixed metal oxides by solution combustion. These are very interesting approaches that give further support to the sustainability of metal precursors used in the SCS. Nevertheless, the utilization of gas capture or NO_x abatement to control NO_x emissions to the atmosphere is still required.

The relatively high decomposition temperatures for alkali and alkaline metal nitrates compared to transition metals are a consequence of the low charge density of the metal cations, and hence their inefficiency to polarize the nitrate anion, favoring a quite strong ionic bonding. Interestingly, this trend is also reflected in the ignition temperatures of the metal nitrate/glycine redox mixture (Kingsley and Pederson, 1993). On the other hand, the covalent metal nitrates for the transition metal cations not only exhibit the effect of charge density, but also a possible back-donation of electrons (Yuvaraj *et al.*, 2003; Jones, 1973) that decrease the N–O bond and consequently the decomposition and reduction temperatures. The thermal decomposition of several first-row transition metal nitrates (i.e., Mn, Co, Ni, Cu, and Zn) supported on alumina occurs at lower temperatures than the unsupported metal nitrates (Małecka *et al.*, 2015), indicating that the presence of support–metal nitrate interaction would bring an additional benefit to these sustainable metal nitrates based on the energy efficiency of the process.

1.4.2. *Organic Compounds as Fuels*

Organic fuels alongside metal nitrates are usually used as redox mixture to produce a sustained combustion reaction on the preparation of bulk catalysts, metal oxide-, or metal-supported catalysts via the SCS approach (González-Cortés and Imbert, 2013; Thoda *et al.*, 2018; Wolf *et al.*, 2019). It is also possible to apply other metal precursors and even a combination of renewable or nonrenewable fuels to produce the redox mixture (Deganello and Tyagi, 2018). To find out the greenness of different fuels based on the emission of greenhouse gases (GHGs), a metric called *gas emissions factor* (GEF) to quantify the content of the anthropogenic gases produced on the SCS of a solid catalyst is defined according to Eq. (1.3). This metric takes into account not only the production of CO_x (CO/CO_2), but also NO_x and any other hazardous gases (i.e., HCNO, HCN, and NH_3).

$$\text{Gas emissions factor (GEF)} = \frac{\text{Total mass of emissions}}{\text{Total mass of fuel in the reactant}}$$

(1.3)

Carbon is usually one of the main components of the organic fuel and the source of carbon dioxide; hence, the calculation of CO_2 emitted on combustion is a straightforward carbon footprint metric to assess the sustainability of the organic fuel. For example, the CO_2 emissions of 100 g of urea produce 73.3 g of CO_2 according to the stoichiometric combustion reaction Eq. (1.4).

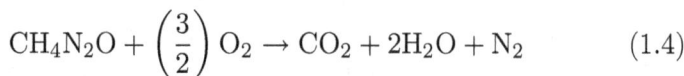

$$CH_4N_2O + \left(\frac{3}{2}\right) O_2 \rightarrow CO_2 + 2H_2O + N_2$$

(1.4)

A similar calculation for CO_2 emissions was carried out for a wide variety of organic compounds/fuels to illustrate the suitability of the GEF metric and the sustainability of these fuels. Note that N-containing gases (i.e., N_2, NO_x, NH_3, and HNO_3) were not taken into account since their concentrations are strongly dependent on the fuel, metal precursor, fuel-to-oxidizer ratio, and also the operation mode to initiate the combustion reaction (Kang *et al.*, 2018; Manukyan *et al.*, 2013). Hence, they need to be experimentally measured to have a reliable total emission factor.

The CO_2 emissions versus the reducing valence for organic compounds with different functional groups (i.e., carboxylic, hydroxyl, carbonyl, amino, and their combinations) are shown in Fig. 1.10. As expected, the CO_2 generated upon the CS rises as the reducing valence of the organic fuel increases because of the higher content of stoichiometric carbon and enthalpy of combustion (González-Cortés and Imbert, 2013).

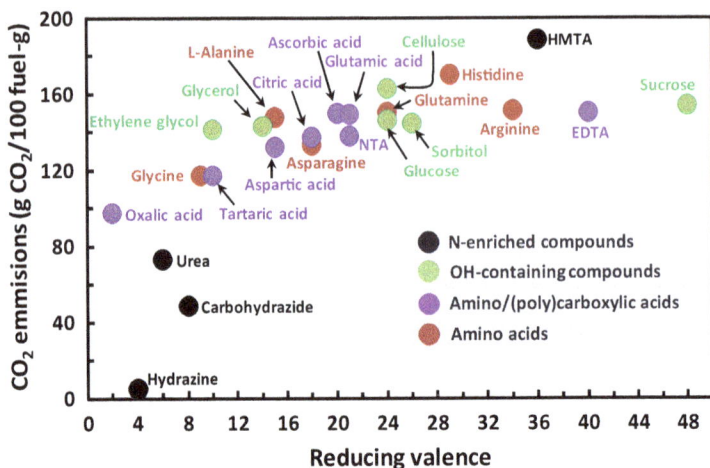

Figure 1.10. Dependence of the carbon dioxide emissions with the reducing valence of a variety of organic fuels used in the SCS.

Hydrazine (N_2H_4) does not produce carbon dioxide, whereas carbohydrazide (CH_6N_4O) and urea (CH_4N_2O) show significantly lower CO_2 emissions than hexamethylenetetramine (HMTA), carboxylic acids, glycols, amino acids, and carbohydrates. However, it is envisaged that these N-containing chemicals with low reducing valence would produce high concentrations of waste nitrogen oxides (NO_x) owing to the oxidation reaction of nitrogen, particularly hydrazine, upon the combustion process because of their relatively high N/C ratio. Furthermore, hydrazine and carbohydrazide are hazardous and toxic substances that would severely limit the greenness of the SCS approach. On the other hand, urea (called carbamide as well) also produces comparatively low emissions of CO_2 and is essentially a nonhazardous and nontoxic substance that boosts the sustainability of CS of catalysts (González-Cortés *et al.*, 2005; Zhao *et al.*, 2019).

Glycine ($C_2H_5NO_2$) is grouped into the series of amino acid-type fuel whose two functional groups (i.e., carboxylic and amino) give to it an amphoteric and zwitterion characteristics in aqueous solution. This organic fuel has been extensively used in the

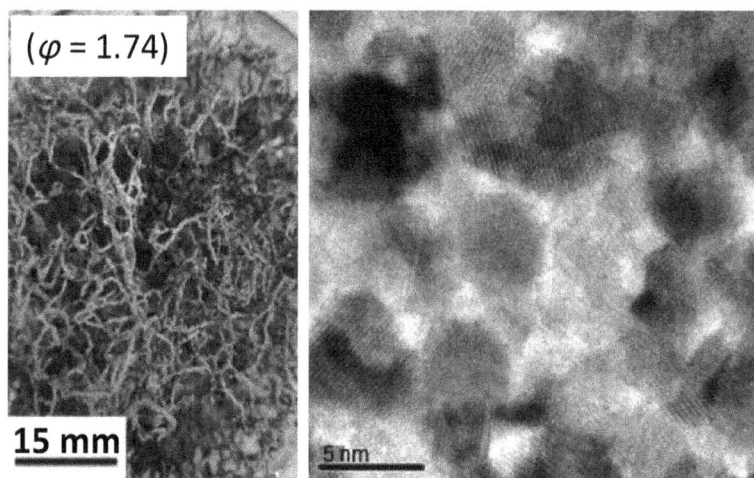

Figure 1.11. The hierarchical-type network of Ni/NiO catalyst. Adapted from (Xanthopoulou *et al.*, 2018).

synthesis of nanomaterials (Fathi *et al.*, 2017; Hwang and Wu, 2004; Khaliullin *et al.*, 2016; Manukyan *et al.*, 2013) and catalysts (Rodriguez-Sulbaran *et al.*, 2018; Xanthopoulou *et al.*, 2018; Zavyalova *et al.*, 2007) because of its potential to tune the oxidizing character of the combustion atmosphere with the fuel-to-oxidizer ratio and to produce three-dimensional hierarchical and dendrite structures as shown in Fig. 1.11. However, this type of fuel is partially limited from the point of view of sustainability because of its relatively high CO_2 (and NO_x) emissions upon the combustion process, see Fig. 1.10. Sustainable methodologies to produce amino acids, particularly glycine, would substantially boost the greenness of the SCS (Uneyama *et al.*, 2017).

Edible carboxylic acids such as citric acid, ascorbic acid, tartaric acid, oxalic acid, among others, and amino polycarboxylic acids are also suitable organic fuels for the synthesis of a variety of materials and catalysts by solution combustion (Chen *et al.*, 2014; Gardey-Merino *et al.*, 2015; Gardey-Merino *et al.*, 2012; González-Cortés *et al.*, 2015; Jian *et al.*, 2014; Pavithra *et al.*, 2017). However, those chemicals are also used as food additives, limiting their potential sustainability for the SCS approach. An eco-friendly and likely carbon neutral approach is the synthesis of nanopowder materials and catalysts using renewable fuels such as cellulose (Ashok *et al.*, 2015; Ashok *et al.*, 2016; Kumar, 2019), sorbitol (Ajamein and Haghighi, 2015; Ajamein *et al.*, 2017), glucose (Li *et al.*, 2016; Srikesh and Nesaraj, 2015), sucrose (Huang *et al.*, 2017; Rodriguez-Sulbaran *et al.*, 2018), and glycerol (Moraes *et al.*, 2015). Another environmentally friendly approach is the combination of a renewable fuel such as cellulose (or cotton) with glycine, a high carbon footprint fuel, to produce mixed metal oxides (Zhou *et al.*, 2008a; Zhou *et al.*, 2008b) and Ni–NiO nanoparticles (Foo *et al.*, 2017). This strategy should also be relevant to the utilization of carboxylic acids and amino polycarboxilic acids under sustainable conditions.

1.4.3. *Water and pH in Solution*

It is estimated that between 80% and 90% of the MI in the production of active pharmaceutical ingredient (API) corresponds to solvents (Constable *et al.*, 2007; Jimenez-Gonzalez *et al.*, 2012). In general, solvents are of great environmental concerns as they are used in large excess relative to the reactants and products; hence, the ideal green reaction would be the synthesis of chemicals without solvents (Cave *et al.*, 2001; Metzger, 1998). Taking advantage of the hygroscopic character and relatively low melting point of metal nitrates, some researchers have used a nearly dry SCS approach (i.e., molten salt CS) to prepare a variety of single- and mixed-metal oxides (Chen and Yao, 2011; Ringuede *et al.*, 2001). In cases where a solvent is absolutely necessary, its rational selection is a very important prerequisite for the design of green chemicals. Water plays an important role in the CS of nanomaterials.

It was recently reported that the water content in nickel nitrate-glycine solution affects the Ni/NiO compositions, crystallite size, and textural properties (i.e., pores size distribution and surface area), which was also reflected in the nature of the surface active sites and hence the catalytic performance in the hydrogenation of maleic acid (Xanthopoulou *et al.*, 2017) as shown in Fig. 1.12.

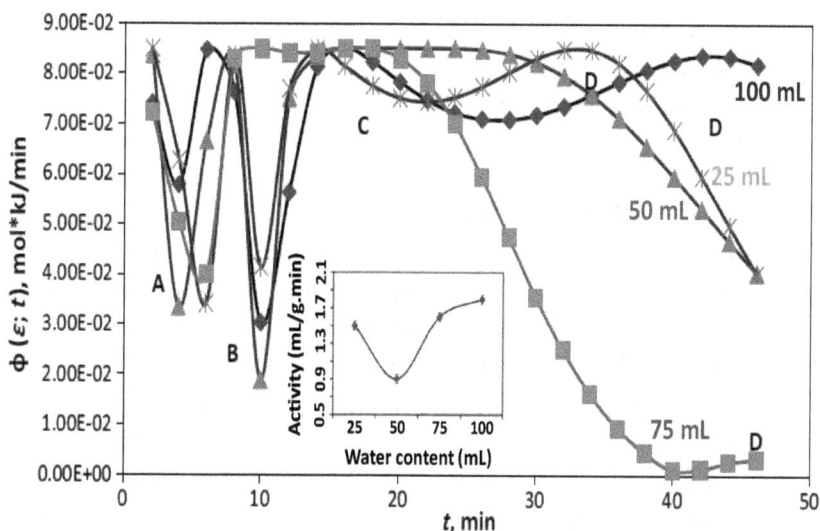

Figure 1.12. The surface distribution of active sites through the local adsorption energies for Ni-based catalysts and dependence of the catalyst performance with water content (inset figure). Adapted from Xanthopoulou *et al.* (2017).

This finding may be associated not only to the number of water and glycine molecules coordinated to Ni^{2+} cation to produce $[Ni(H_2O)_n(Gly)_x]^{2+}$, but also to the bidentate ligand character of glycine when the redox gel is treated at various temperatures. This can affect the maximum temperature of the redox mixture combustion (Manukyan *et al.*, 2013). A strong influence of the initial concentration of the metal precursors over the structural and morphological properties of $(Co, Ni)Cr_2O_4$ pigments synthesized by CS was also observed (Gilabert *et al.*, 2017).

Within the context of the SCS of nanostructured materials or catalysts, it is expected that at least 50% of the MI is associated with solvents as they are usually used to fully dissolve the reactants (fuels and oxidizers) and also for absorbing and transferring heat on the thermal treatment. Fortunately, water is the most common solvent used in the SCS method because of the high solubility of most metal precursors, usually metal nitrates, and polar organic compounds. Furthermore, water is environmentally friendly because of its nontoxic and nonflammable character; it is also economically accessible because of its low price and abundant reserves. Indeed, among 51 solvents assessed, water is on the top of the ranking of greener or recommended solvents based on health, safety, and environment criteria (Table 1.2).

Another factor related to aqueous solution is the pH, which is particularly relevant for enhancing the fuel–oxidizer interaction and hence the formation of gel-type network (Komova *et al.*, 2016; Rodriguez-Sulbaran *et al.*, 2018; Wu *et al.*, 2004). The progressive formation of cation-oxygen bonds and subsequent release of H^+ ions can be controlled by the addition of ammonium hydroxide solution to the redox solution (Deganello and Tyagi, 2018; Kim *et al.*, 2005). The addition of a base (or an acid) into the redox solution can certainly affect the crystalline structure, microstructure, morphology, and texture of the synthesized material (Jaimeewong *et al.*, 2016; Lwin *et al.*, 2015; Pourgolmohammad *et al.*, 2017a). However, it also contributes to the generation of further emissions and potential health and safety issues because of the large concentrations of usually strong acid/base substances and the potential increase of the explosion character of the combustion reaction. A greener approach would be the utilization of the optimal stoichiometric composition of the redox mixture (fuel-to-oxidizer ratio) and the adjustment of the water content (or mixture of solvents) and type of fuel (or mixture of fuels) to obtain the required structural, morphological, and textural properties of the nanostructured catalyst or material.

1.4.4. *Ignition Modality of the Redox Mixture*

A major green feature of the SCS method is its energy efficiency, as once the redox mixture achieves the ignition temperature a self-sustained exothermic reaction takes place for a short, and usually sufficient, period of time to decompose the entire metal-organic-nitrate matrix. The final product is usually a nanomaterial with a large specific surface area as a consequence of the large amount of gases produced during the synthesis process and the rapid cooling process. The combustion reaction releases large amount of heat that can also be exploited by fueling the process to ignite fresh redox mixture and hence enhance the sustainability of the SCS process.

Table 1.2. Sustainability ranking of a variety of solvents.

Recommended	Recommended or problematic?	Problematic	Problematic or hazardous?	Hazardous	Highly hazardous
Water	Methanol	2-Methyltetrahydrofuran	Methyl tert-butyl ether	Diisopropylether	Diethylether
Ethanol	tert-Butyl alcohol	Heptane	Tetrahydrofuran	1,4-Dioxane	Benzene
2-Propanol	Benzyl alcohol	Methylcyclohexane	Cyclohexane	Dimethyl ether	Chloroform
1-Butanol	Ethylene glycol	Toluene	Dichloromethane	Pentane	Carbon tetrachloride
Ethyl acetate	Acetone	Xylenes	Formic acid	Hexane	Dichloroethane
2-Propyl acetate	Butanone	Chlorobenzene	Pyridine	Dimethylformamide	Nitromethane
n-Butyl acetate	4-Methyl-2-pentanone	Acetonitrile		N,N-Dimethylacetamide	
Anisole	Cyclohexanone	N,N′-Dimethylpropyleneurea		1-Methyl-2-pyrrolidone	
Sulfolane	Methyl acetate	Dimethyl sulfoxide		Methoxy ethanol	
	Acetic acid			Triethanolamine	
	Acetic anhydride				

Adapted from Prat *et al.* (2014).

When the redox mixture is uniformly heated (i.e., volume combustion mode) the initiation of the combustion reaction can be carried out through controlled and uncontrolled thermal treatments. The major difference between these operation modes lies in the heating rate that the redox mixture is ignited to induce the combustion reaction (González-Cortés and Imbert, 2013). The initiation process is usually carried out by conventional heating through hot plate and, most commonly, muffle furnaces (Deganello and Tyagi, 2018), particularly for the synthesis of solid catalysts.

Innovative initiation approaches such as microwave energy (Rodriguez-Sulbaran *et al.*, 2018) and more recently the cavitation effect of ultrasound applied to the synthesis of porous and nano-sized $Na_3V_2(PO_4)_3/C$ composites (Chen *et al.*, 2017) are emerging alternatives to enhance the greenness of the SCS method. The selective and rapid superheating over dielectric materials through the microwave transfer energy can occur by the ionic conduction, dipolar polarization, and interfacial polarization mechanisms, as shown in Fig. 1.13. The ionic conduction occurs in charged molecules and ionic solutions because of the interaction between the oscillating electric field component of the microwave and the ions. As a result, the ions move back and forth while they also collide each other to produce local overheating. The dipolar polarization mechanism, on the other hand, takes place when molecules with a permanent dipole moment are exposed to microwave irradiation to excite

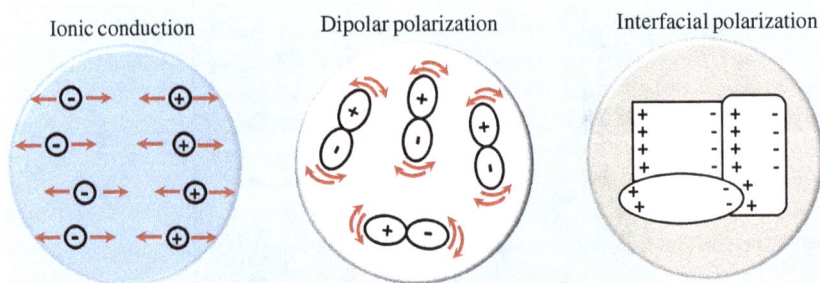

Figure 1.13. Microwave transfer energy through ionic conduction, dipolar polarization, and interfacial polarization mechanisms. Adapted from Dakin (2006).

configurational modes. The nano-domain of polar molecules tries to align with the oscillating applied electric field; the relaxation process and the continual reorientation produce friction between molecules and thereby selective superheating. The interfacial polarization is another relaxation mechanism, in which the heterogeneity of different dielectric interfaces produces transient charge separation and hence localized overheating when an oscillating field is applied (Aldawsari, 2016; Dudley *et al.*, 2015; Mishra and Sharma, 2016; Stuerga, 2006; Dakin, 2006). These mechanisms operate on the SCS of nanomaterials when microwave energy ignites the redox mixture. This is a consequence of the selective absorption of microwave irradiation by polar molecules (i.e., solvent and organic fuel) and lossy compounds, whose dielectric loss can widely vary within a series of materials as shown in Fig. 1.14.

This is often reflected in a higher heating rate and temperature in comparison with conventional heating, leading to the reduction of reaction time because of the increase of reaction rate. The positive effect of microwave-assisted SCS over the morphology, textural properties, and crystallinity of nanostructured catalysts was reported for methanol steam reforming over $CuO-ZnO-Al_2O_3$ (Ajamein and Haghighi, 2016) and for $Ni-Al_2O_3$ catalysts for slurry-phase CO

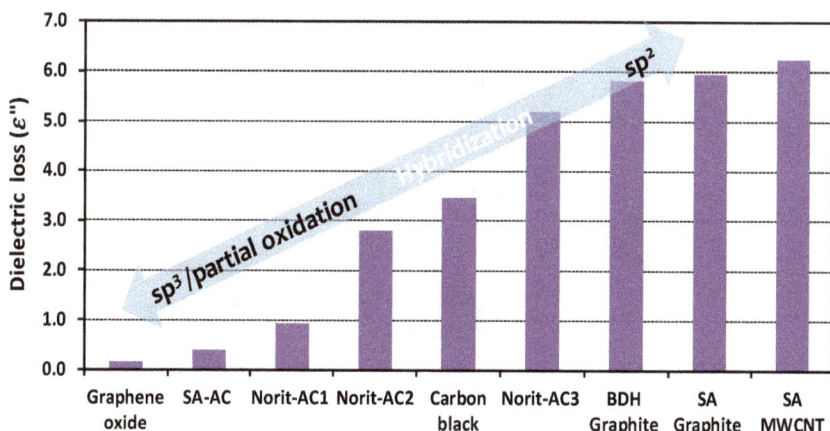

Figure 1.14. Dielectric loss for a variety of carbon-based materials.

methanation (Gao *et al.*, 2016). A comparative study of the effect of the ignition source (i.e., furnace and microwave) over the synthesis of $La_{0.6}Sr_{0.4}Ni_yAl_{1-y}O_3$ perovskite-type catalyst precursors and the catalytic performance for the dry reforming of methane (DRM) did not show an important influence of the initiation mode over the catalyst properties. However, the catalyst composition and type of fuel significantly affected the textural properties and catalytic performances of the LaSrNiAl perovskite-type structures as shown in Fig. 1.15. Nevertheless, the use of microwaves markedly reduced the time of thermal treatment for the catalyst synthesis (Rodriguez-Sulbaran *et al.*, 2018). This particular advantage alongside the utilization of renewable fuels significantly enhances the greenness of the SCS method (Azizi *et al.*, 2017).

According to Kingsley and Pederson (1993), the ignition temperature of metal nitrate/glycine mixture increases in the following order: transition metal nitrates < rare earth nitrates < alkaline earth nitrates, because of different degrees of interaction between the cation, glycine, and nitrate. In fact, it is well established that

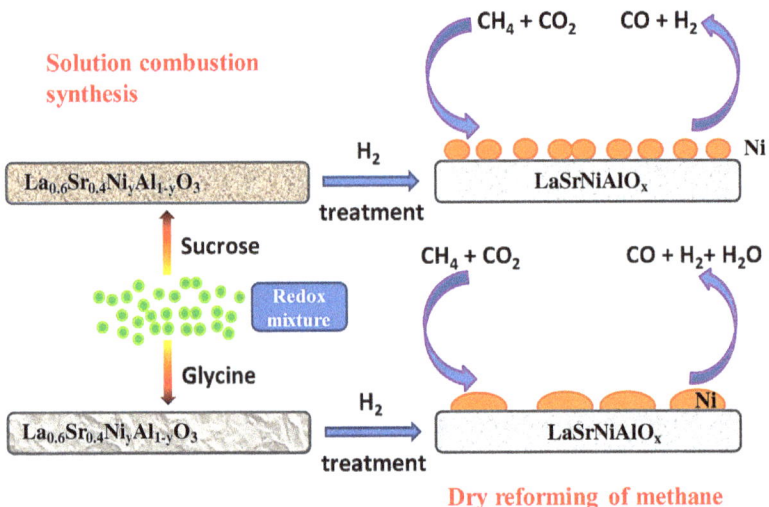

Figure 1.15. Effect of sucrose and glycine over the textural properties and catalytic performance of LaSrNiAl perovskite-type structures. Adapted from Rodriguez-Sulbaran *et al.* (2018).

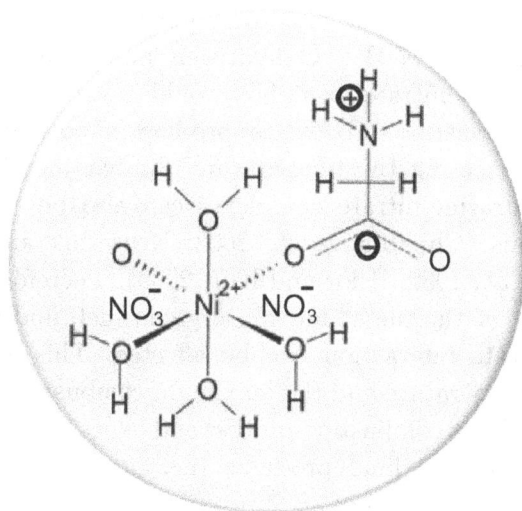

Figure 1.16. Main structural element for the Ni–glycine–nitrate complex. Adapted from Fleck and Bohatý (2005b).

glycine can coordinate to various cations at room temperature to produce crystalline structures of metal–glycine–nitrate complex (Davies *et al.*, 1992; Fleck and Bohatý, 2005a, 2005b). Figure 1.16 shows the crystalline structure for Ni–glycine–nitrate complex (i.e., catena-Poly[[[tetraaquanickel(II)]-1-glycine-κ^2O:O'] dinitrate]), which is isostructural with its magnesium and cobalt analogous (Fleck and Bohatý, 2005a), synthesized at room temperature from a nickel nitrate–glycine aqueous solution.

The main structural feature of this compound is the octahedral coordination of Ni^{2+} through six O atoms, four of these O atoms belong to water molecules and two, in a cis configuration, belong to the carboxylate groups of glycine molecules. Furthermore, the N-terminal ends of the glycine molecules in their zwitterionic form produce N–H$\cdots$O hydrogen bonds to the O atoms of the nitrate anions located in the interstices between the chains. The O atoms of the nitrates act as acceptors for the hydrogen bonds generated by the amino groups and the water molecules. The bifunctionality of glycine facilitates the formation of dendrite-type morphology in the synthesized metal oxides (Thoda *et al.*, 2019).

It is envisaged that the nano-proximity of the amino group to the nitrate through N–H···O hydrogen bond into the structure of the Ni–glycine–nitrate complex would favor its ignition and subsequent combustion process to produce nanostructured nickel oxide when increasing the temperature. Similar metal-urea-nitrate and metal-hydrazine-nitrate complexes can also be synthesized at mild conditions (Chhabra *et al.*, 2003; Krawczuk and Stadnicka, 2007; Patil *et al.*, 1982; Prior and Kift, 2009). Therefore, depending on the nature of the metal nitrate, organic fuel, and solution pH, metal-fuel-nitrate interaction can be affected. This is reflected in the ignition temperature and the maximum combustion temperature achieved during the combustion process and hence the microstructure characteristics of the final products (i.e., material and catalyst). Interestingly, the selection of an appropriate fuel and fuel-to-oxidizer ratio (Kumar *et al.*, 2011; Wolf *et al.*, 2019) or a fuel-oxidizer mixture together with ignition source (i.e., microwave irradiation) and solution pH (Khort *et al.*, 2017; Podbolotov *et al.*, 2017) can produce metal nanopowders through one-pot synthesis, indicating that the sustainability of CS can be enhanced through process intensification. This approach is particularly relevant for the synthesis of metal catalysts, which are usually obtained after a reduction step of the metal oxide catalyst precursors.

1.5. A Holistic Perspective of Sustainable SCS Processes

The development of clean, safe, and environmentally friendly SCS technology depends on the utilization of less toxic metal precursors and organic fuels, as well as water as a environmentally benign solvent. It is also important to minimize the number of reagents and synthetic steps through process intensification and to decrease the energy consumption alongside the generation of by-products and waste emissions. The main parameters that individually and collectively affect the greenness of the SCS of nanostructured catalysts are shown in Fig. 1.17. It is worth remarking that the effects of the single factor over the structure, microstructure, and morphology of

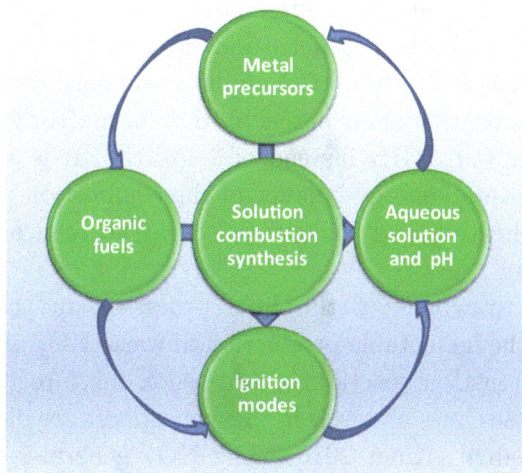

Figure 1.17. Network of factors that can affect the greenness of the SCS methods.

nanostructured materials are widely covered in existing literature, as discussed in the previous sections. The combination of two or several factors could bring possible cooperative effects or synergy that would boost the greenness of CS.

It is anticipated that microwave energy as ignition mode and the utilization of sustainable organic fuels and metal precursors alongside small content (or free) of water, to produce the cation-organic-nitrate redox mixture, would enhance the greenness of the SCS process. Of course, all these parameters need to be properly tuned to synthesize catalysts (or materials) with optimal performance (e.g., activity, selectivity and stability for solid catalysts), while minimizing environmental and health impacts through economically viable technology. The microwave-assisted SCS is a suitable greener approach to produce nanostructured catalysts because of the shortness of the thermal treatment, hence enhancing the energy efficiency (Frikha *et al.*, 2019; Rodriguez-Sulbaran *et al.*, 2018). Nevertheless, microwave (and ultrasound) energy applied to the SCS is still in its infancy and its full potential to control structure, microstructure, and morphology is not yet understood in terms of the dielectric properties of the redox mixture, reaction and heat mechanisms, and the optimization of reaction parameters. Furthermore, the relatively low microwave penetration into large volume material-containing reactors for industrial applications needs to be overcome through the utilization of novel reactor designs and/or low microwave frequency (i.e., 0.915 GHz instead of 2.45 GHz). It is also necessary to assess the overall energy balance of the combustion process under conventional heating and microwave energy to realize its potential applications.

A major drawback of the SCS processes, in the context of greenness, is the ineluctable production of waste NO_x and CO_x gases from the combustion reaction of the redox mixture (i.e., fuel and oxidizer), whose emission levels in the atmosphere are directly related to climate change (King, 2016). The NO_x generates mainly from nitrates as metal precursors; however, fuels such as urea, hydrazine, and amino acids can also contribute to NO_x emissions upon the combustion process. The production of GHGs can be substantially neutralized using renewable organic fuels such as cellulose (Foo *et al.*, 2017), sucrose (Amarilla *et al.*, 2011), and glucose (Li *et al.*, 2016), since their carbon contribution can be considered partially neutral because extra fossil fuel energy is required during the lifecycle of biomass. More importantly, the implementation of carbon capture and storage (CCS), NO_x abatement, and even NO_x capture (Bui *et al.*, 2018; Shah *et al.*, 2012; Silas *et al.*, 2019) can also contribute to reducing GHG and even produce a negative carbon footprint, hence markedly enhancing the sustainability and greenness of CS.

Health and safety matters are also important principles for green SCS methods not only due to the generation of hazardous gases but also, and perhaps most importantly, due to the rapid and abrupt release of a large amount of gases and heat through a violent and exothermic combustion reaction (Deganello and Tyagi, 2018). To minimize the potential explosion risk of the combustion reaction, several recommendations need to be followed: (1) The thermal treatment should be carried out under controlled condition (i.e., low heating rate) (González-Cortés and Imbert, 2013). (2) A large volume reactor vessel should be used, relative to the volume of synthesized solid, that allows gases to rapidly release to the atmosphere to minimize the increase of the pressure upon the combustion reaction. (3) The reaction needs to be made in a ventilation enclosure to reduce the danger of exposing the researcher to hazardous gases such as CO, NH_3, HNO_3, and NO_x. Another option to minimize risk upon the combustion process can be a self-sustained smouldering combustion reaction, in which the selection of an appropriate fuel with high reducing valence (e.g., polycarboxilic acids or amino polycarboxilic acids) or a sustainable fuel is paramount to control the explosive character of the combustion reaction. Finally, the one-pot SCS of metal materials (catalysts) (Khort *et al.*, 2017; Podbolotov *et al.*, 2017; Trusov *et al.*, 2016) also requires extra precautions because of the high pyrophoric character of noble and base metals that rapidly heat up and become an ignition source for flammable materials.

In summary, the greenness and sustainability of the SCS methods and the catalytic performance (i.e., activity, selectivity and stability) can be enhanced when a rational selection of synthesis parameters such as: (1) low decomposition and reduction temperature of abundant metal precursors; (2) high solubility in water; (3) neutral acidity; (4) low tendency to contaminate the final catalyst or material; (5) minimization of noxious emissions during the combustion process (i.e., low gas emission factor); (6) utilization of waste-derived precursors and sustainable organic fuels; (7) ignition of redox mixture with microwave (ultrasound); (8) high reducing valence of the fuel, particularly polycarboxilic acids; and amino polycarboxilic acids; and (9) process intensification, among others, are properly tuned.

1.6. Conclusion and Outlook

It is evident that the emerging field of green SCS of nanostructured catalysts needs to overcome some challenges from the viewpoint of sustainability to achieve the maximum performance and benefit from solution combustion method, while minimizing its impact not only on human health and the environment, but also producing an economically viable technology. In this context, the 12 principles of green chemistry, applied to the synthesis of catalysts and materials, provide a framework for designing advanced nanostructured materials and develop novel methods able to face the future demand for energy generation and supply. Several factors have been identified that should be taken into account to design greener nanostructured catalysts and materials through the SCS process:

1. *Metal precursor.* It needs to be obviously linked not only to the performance of the catalyst (i.e., activity, selectivity, and stability) or particular application of the material, but also to the decomposition temperature of the metal nitrate and the metal abundance in the earth's crust. Particularly relevant is the first-row transition metals (or base metals) whose relatively low decomposition temperatures for their metal nitrates and high abundance; alongside minimal safety concerns, low cost and global availability make them the ideal metals for designing green catalysts and materials.

2. *Organic fuel.* It is a critical component for the greenness and sustainability of the SCS of catalysts and materials because of its relatively low AE. This is a consequence of the large generation of waste CO_x and NO_x emissions during the combustion reaction of the fuel-oxidizer redox mixture. These gases can be substantially neutralized through the utilization of sustainable organic fuels, which are abundant in nature, accessible in cost and safe as a handling reagent. Furthermore, the possible implementation of CCS, NO_x abatement, and NO_x capture can minimize emissions and even favor a negative carbon footprint, markedly enhancing the greenness of CS.

3. *Ignition mode and aqueous solution.* Two relatively new approaches based on the utilization of microwave and ultrasound energy have been applied to enhance the greenness of the SCS method. They may be able to increase the energy efficiency of the combustion process as a consequence of the reduction of the thermal treatment period and hence the energy consumption. However, future works should be focused on determining experimentally the energy efficiency of this process. Within the sustainability context, water is the most common solvent used in the

SCS method because of its high polarity and dielectric constant. It is also the most environmentally friendly solvent with nontoxic and nonflammable properties; it is also economically accessible because of its low price and abundant reserves. A greener method is the synthesis of nanostructured materials without solvent where the hygroscopic character and relatively low melting point of metal nitrates can facilitate the homogenization of the redox mixture.

Despite the progress achieved in the SCS method as described in this chapter, there are still considerable challenges in this subject that need to be addressed to overcome the current limitations. To pursue further advances not only in the fundamentals of CS, but also in the development of greener approaches, several research fields would have a significant impact on future directions as described below.

There is a need to understand the effect of synthesis parameters over the size, morphology, and composition of the nanostructured materials. Currently, there has been very little progress in the mechanistic studies and systematic modelling to gain insight into the dynamics of the SCS process both in the gas phase and the solid phase. There is a lack of methodology to control not only the size within the nanometer and micrometer scales, but also the shapes of various geometries for nanostructured materials. An in-depth understanding of the controllability and flexibility of these factors would lead to further advances in the control of the morphology of nanomaterials. This particular challenge can be met through the combination of developing new synthetic approaches to control the exothermicity of the combustion reaction and *in-situ* advanced analytical techniques to monitor the combustion reaction under real operation time.

To develop greener SCS approaches, the nanostructure- and microstructure-controlling agents (i.e., fuels and soft/hard templates) from natural resources are used to neutralize carbon footprint and minimize the use of hazardous chemicals. For instance, renewable raw materials such as proteins and deoxyribonucleic acids (DNAs) are usually available and abundant in nature and also potentially useful fuels for SCS. It will be particularly relevant to identify natural products that can play a bifunctional role as a fuel and structure-directing agent to produce outstanding nanostructured catalysts and materials. Currently, although biomass-derived fuels have been used to produce nanostructured materials, there is still a lack of understanding of the influence of several metrics (pH, solvent, metal precursor, fuel-to-oxidizer ratio, etc.) over the synthesis mechanism and hence the size, morphology, and composition of the nanostructured materials. Hence, further investigation in this field would be worthwhile.

The intensification of the SCS of nanostructured materials through suitable cascade (i.e., tandem or domino) reactions is an interesting strategy to develop advanced materials. One-pot synthesis involving several consecutive reactions decreases the energy and raw material consumption through the reduction of steps such as the separation and purification of intermediates, giving further benefit to the greenness of the SCS process. Currently, little is known about the mechanistic details of the transformation of metal nitrate to metal oxide and the subsequent formation, and possible autocatalytic effect, of nano-crystalline metal particles. This approach is particularly relevant for the synthesis of metal catalysts, which are usually obtained through two stages (i.e., synthesis of metal oxides and its subsequent reduction step). It is worth also exploring the potential of the SCS to make mixed-metal carbide and sulfide catalysts in one-pot synthesis.

There is a need to use appropriate metrics to assess the greenness and sustainability of different approaches of CS. Mass-based metrics considering the stoichiometry of chemical reactions need to be supplemented by metrics that measure not only the environmental impact of the raw materials (i.e., fuels and metal precursors), product, and waste, but also the energy efficiency, economic, and societal metrics within a broader life cycle analysis. Although this methodology is more suitable for technology in an advanced stage of development, it can give further insight into the key metrics to be considered in the development of new, greener approaches for the SCS methods.

Looking at the road to a green and sustainable future, increasing awareness of the importance of greenness and sustainability has emerged into every field of chemistry guided by the 12 principles. This holistic approach applied to the SCS of nanostructured catalysts and materials clearly reveals that this synthesis method is still in its early stages and further research is necessary to enhance the current methods, which can be supplemented by the utilization of renewable energy (hydroelectricity, wind energy, wave power, geothermal energy, or bio-energy). Regardless of the green approach applied, it is obviously expected that green chemistry will pave the way for developing advanced technologies that meet the standards of greenness and sustainability to fuel our society and protect our environment.

Acknowledgments

The authors thank all the coauthors cited in this chapter and the University of Oxford for generous funding and continual support of this multidisciplinary research.

References

Ajamein, H., & Haghighi, M. (2015). Effect of sorbitol/oxidizer ratio on microwave assisted solution combustion synthesis of copper based nanocatalyst for fuel cell grade hydrogen production. *Iranian J. Hyd. & Fuel Cell*, *4*, 227–240.

Ajamein, H., & Haghighi, M. (2016). On the microwave enhanced combustion synthesis of CuO–ZnO–Al$_2$O$_3$ nanocatalyst used in methanol steam reforming for fuel cell grade hydrogen production: Effect of microwave irradiation and fuel ratio. *Energy Convers. Manag.*, *118*, 231–242.

Ajamein, H., Haghighi, M., & Alaei, S. (2017). The role of various fuels on microwave-enhanced combustion synthesis of CuO/ZnO/Al$_2$O$_3$ nanocatalyst used in hydrogen production via methanol steam reforming. *Energy Convers. Manag.*, *137*, 61–73.

Aldawsari, A. M. (2016). *Dielectric Measurements and Catalytic Cracking of Heavy oils Using Advanced Microwave Technologies*. Oxford: Oxford University.

Amarilla, J. M., Rojas, R. M., & Rojo, J. M. (2011). Understanding the sucrose-assisted combustion method: Effects of the atmosphere and fuel amount on the synthesis and electrochemical performances of LiNi$_{0.5}$Mn$_{1.5}$O$_4$ spinel. *J. Power Sources*, *196*, 5951–5959.

Anastas, P. T., & Kirchhoff, M. M. (2002). Origins, current status, and future challenges of green chemistry. *Acc. Chem. Res., 35,* 686–693.

Anastas, P. T., Kirchhoff, M. M., & Williamson, T. C. (2001). Catalysis as a foundational pillar of green chemistry. *Appl. Catal. A: Gen., 221,* 3–13.

Aruna, S. T., & Mukasyan, A. S. (2008). Combustion synthesis and nanomaterials. *Curr. Opin. Solid State Mater. Sci., 12,* 44–50.

Ashok, A., Kumar, A., Bhosale, R. R., Ali, M., Saleh, H., & van den Broeke, L. J. P. (2015). Cellulose assisted combustion synthesis of porous Cu–Ni nanopowders. *RSC Adv., 5,* 28703–28712.

Ashok, A., Kumar, A., Bhosale, R. R., Saleh, M. A. H., Ghosh, U. K., Al-Marri, M., . . . Tarlochan, F. (2016). Cobalt oxide nanopowder synthesis using cellulose assisted combustion technique. *Ceram. Int., 42,* 12771–12777.

Azapagic, A., & Perdan, S. (2000). Indicators of sustainable development for industry: A general framework. *Trans. Inst. Chem. Eng., 78,* 243–261.

Azizi, S., Mohamad, R., & Shahri, M. M. (2017). Green microwave-assisted combustion synthesis of zinc oxide nanoparticles with Citrullus colocynthis (L.) Schrad: Characterization and biomedical applications. *Molecules, 22,* 301.

Bera, P., & Aruna, S. T. (2018). Solution combustion synthesis, characterization, and catalytic properties of oxide materials. In M. Van de Voorde & B. Sels (Eds.), *Nanotechnology in Catalysis: Applications in the Chemical Industry, Energy Development, and Environment Protection* (pp. 91–118). Weinheim: Wiley-VCH Verlag GmbH & Co. KGaA.

Bravo-Suarez, J. J., Chaudhari, R. V., & Subramaniam, B. (2013). Design of heterogeneous catalysts for fuels and chemicals processing: An overview. In J. J. Bravo-Suarez, M. K. Kidder, & V. Schwartz (Eds.), *Novel Materials for Catalysis and Fuels Processing* (pp. 1–68). Washington: ACS Symposium Series.

Brockner, W., Ehrhardt, C., & Gjikaj, M. (2007). Thermal decomposition of nickel nitrate hexahydrate, $Ni(NO_3)_2 \cdot 6H_2O$, in comparison to $Co(NO_3)_2 \cdot 6H_2O$ and $Ca(NO_3)_2 \cdot 4H_2O$. *Thermochim. Acta, 456,* 64–68.

Brundtland, G. H. (1987). World commission on environment and development. In *Our Common Future.* Oxford: Oxford University Press.

Bui, M., Adjiman, C. S., Bardow, A., Anthony, E. J., Boston, A., Brown, S., . . . Mac Dowell, N. (2018). Carbon capture and storage (CCS): The way forward. *Energy Environ. Sci., 11,* 1062–1176.

Burgos-Montes, O., Moreno, R., Colomer, M. T., & Fariñas, J. C. (2006). Influence of combustion aids on suspension combustion synthesis of mullite powders. *J. Eur. Ceram. Soc., 26,* 3365–3372.

Butkute, S., Gaigalas, E., Beganskiene, A., Ivanauskas, F., Ramanauskas, R., & Kareiva, A. (2018). Sol-gel combustion synthesis of high-quality chromium-doped mixed-metal garnets $Y_3Ga_5O_{12}$ and $Gd_3Sc_2Ga_3O_{12}$. *J. Alloy Compd., 739,* 504–509.

Carlos, E., Martnis, R., Fortunato, E., & Branquinho, R. (2020). Solution combustion synthesis: Towards a sustainable approach for metal oxides. *Chem. Eur. J.* https://doi.org/10.1002/chem.202000678.

Calvo-Flores, F. G. (2009). Sustainable chemistry metrics. *ChemSusChem*, *2*, 905–919.

Cave, G. W. V., Raston, C. L., & Scott, J. L. (2001). Recent advances in solventless organic reactions: Towards benign synthesis with remarkable versatility. *Chem. Commun.*, *2001*, 2159–2169.

Chandradass, J., Balasubramanian, M., & Kim, K.-H. (2010). Emulsion combustion synthesis. In M. Lackner (Ed.), *Combustion Synthesis: Novel Routes to Novel Materials* (pp. 25–32). Vienna: Bentham Science Publishers.

Chen, Q., Liu, Q., Chu, X., Zhang, Y., Yan, Y., Xue, L., & Zhang, W. (2017). Ultrasonic-assisted solution combustion synthesis of porous $Na_3V_2(PO_4)_3/C$: Formation mechanism and sodium storage performance. *J. Nanopart. Res.*, *9*, 146–155.

Chen, R., & Yao, W. (2011). Molten salt combustion synthesis of $LiMn_2O_4$ at 600°C: The effect of calcination time. *Adv. Mater. Res.*, *230–232*, 457–460.

Chen, T., Lin, H., Cao, Q., & Huang, Z. (2014). Solution combustion synthesis of $Ti_{0.75}Ce_{0.15}Cu_{0.05}W_{0.05}O_2$-$\lambda$ for low temperature selective catalytic reduction of NO. *RSC Adv.*, *4*, 63909–63916.

Chhabra, J. S., Talawar, M. B., Makashir, P. S., Asthana, S. N., & Singh, H. (2003). Synthesis, characterization and thermal studies of (Ni/Co) metal salts of hydrazine: Potential initiatory compounds. *J. Hazard. Mater.*, *99*, 225–239.

Choudhary, R., Koppala, S., & Swamiappan, S. (2015). Bioactivity studies of calcium magnesium silicate prepared from eggshell waste by sol–gel combustion synthesis. *J. Asian Ceram. Soc.*, *3*, 173–177.

Colomer, M. T., Fumo, D. A., Jurado, J. R., & Segadães, A. M. (1999). Non-stoichiometric La$(1 - x)$NiO$(3 - \delta)$ perovskites produced by combustion synthesis. *J. Mater. Chem.*, *9*, 2505–2510.

Constable, D. J. C., Jimenez-Gonzalez, C., & Henderson, R. K. (2007). Perspective on solvent use in the pharmaceutical industry. *Org. Process Res. Dev.*, *11*, 133–137.

Constable, J. C., Curzons, A. D., & Cunningham, V. L. (2002). Metrics to 'green' chemistry — which are the best? *Green Chem.*, *4*, 521–527.

Curran, M. A. (2013). Life cycle assessment: A review of the methodology and its application to sustainability. *Curr. Opin. Chem. Eng.*, *2*, 273–277.

Curzons, D., Constable, D. J. C., Mortimer, D. N., & Cunningham, V. L. (2001). So you think your process is green, how do you know? *Green Chem.*, *3*, 1–6.

Dakin, T. W. (2006). Conduction and polarization mechanisms and trends in dielectrics. *IEEE Electr. Insul. M.*, *22*, 11–28.

Danks, A. E., Hall, S. R., & Schnepp, Z. (2016). The evolution of 'sol–gel' chemistry as a technique for materials synthesis. *Mater. Horiz.*, *3*, 91–112.

Davies, H. O., Gillard, R. D., Hursthouse, M. B., Mazid, M. A., & Williams, P. A. (1992). Boussingault's mixed copper(ii) glycinate nitrate. *J. Chem. Soc., Chem. Commun.*, *1992*, 226–227.

de Jong, K. P. (2009a). Deposition precipitation. In K. P. de Jong (Ed.), *Synthesis of Solid Catalysts* (pp. 111–132). Weinheim: Wiley-VCH Verlag GmbH & Co. KGaA.

de Jong, K. P. (2009b). General aspects. In K. P. de Jong (Ed.), *Synthesis of Solid Catalysts* (pp. 3–10). Weinheim: Wiley-VCH.

Deganello, F., Joshi, M., Liotta, L. F., La Parola, V., Marcì, G., & Pantaleo, G. (2019). Sustainable recycling of insoluble rust waste for the synthesis of iron-containing perovskite-type catalysts. *ACS Omega, 4*, 6994–7004.

Deganello, F., & Tyagi, A. K. (2018). Solution combustion synthesis, energy and environment: Best parameters for better materials. *Prog. Cryst. Growth Charact. Mater., 64*, 23–61.

Deshpande, K., Mukasyan, A., & Varma, A. (2004). Direct synthesis of iron oxide nanopowders by the combustion approach: Reaction mechanism and properties. *Chem. Mater., 16*, 4896–4904.

Deutschmann, O., Knözinger, H., Kochloefl, K., & Turek, T. (2009). Heterogeneous catalysis and solid catalysts. In *Ullmann's Encyclopedia of Industrial Chemistry*. Weinheim: Wiley-VCH Verlag GmbH & Co. KGaA.

Dudley, G. B., Richert, R., & Stiegman, A. E. (2015). On the existence of and mechanism for microwave-specific reaction rate enhancement. *Chem. Sci., 6*, 2144–2152.

Fathi, H., Masoudpanah, S. M., Alamolhoda, S., & Parnianfar, H. (2017). Effect of fuel type on the microstructure and magnetic properties of solution combusted Fe_3O_4 powders. *Ceram. Int., 43*, 7448–7453.

Fleck, M., & Bohatý, L. (2005a). Catena-poly[[[tetraaquamagnesium(II)]-l-glycine-$\kappa 2O{:}O'$] dinitrate]. *Acta Crystallogr. E, 61*, m1887–m1889.

Fleck, M., & Bohatý, L. (2005b). Catena-poly[[[tetraaquanickel(II)]-l-glycine-$\kappa 2O{:}O'$] dinitrate]. *Acta Crystallogr. E, 61*, m1890–m1893.

Foo, Y.-T., Chan, J. E.-M., Ngoh, G.-C., Abdullah, A. Z., Horri, B. A., & Salamatinia, B. (2017). Synthesis and characterization of NiO and Ni nanoparticles using Nanocrystalline Cellulose (NCC) as a template. *Ceram. Int., 43*, 16331–16339.

Frikha, K., Limousy, L., Bouaziz, J., Bennici, S., Chaari, K., & Jeguirim, M. (2019a). Elaboration of alumina-based materials by solution combustion synthesis: A review. *C. R. Chimie, 22*, 206–219.

Frikha, K., Limousy, L., Bouaziz, J., Chaari, K., Josien, L., Nouali, H., … Bennici, S. (2019b). Binary oxides prepared by microwave-assisted solution combustion: Synthesis, characterization and catalytic activity. *Materials, 12*, 910.

Gabal, M. A., Al-Harthy, E. A., Al-Angari, Y. M., Salam, M. A., & Asiri, A. M. (2016). Synthesis, characterization and magnetic properties of MWCNTs decorated with Zn-substituted $MnFe_2O_4$ nanoparticles using waste batteries extract. *J. Magn. Magn. Mater., 407*, 175–181.

Gao, Y., Meng, F., Li, X., Wen, J. Z., & Li, Z. (2016). Factors controlling nanosized $Ni–Al_2O_3$ catalysts synthesized by solution combustion for slurry phase CO methanation: The ratio of reducing valences to oxidizing valences in redox systems. *Catal. Sci. Technol., 6*, 7800–7811.

Gardey-Merino, M. C., Arreche, R., Lassa, M. S., Lascalea, G. E., Estrella, A., & Rodriguez, M. E. (2015). Synthesis of Co-Cu-Mn oxides deploying different fuels. *Revista Materia, 20*, 779–786.

Gardey-Merino, M. C., Palermo, M., Belda, R., Fernandez de Rapp, M. E., Lascalea, G. E., & Vazquez, P. G. (2012). Combustion synthesis of Co_3O_4 nanoparticles: Fuel ratio effect on the physical properties of the resulting powders. *Procedia Mater. Sci., 1*, 588–593.

Gilabert, J., Palacios, M. D., Sanz, V., & Mestre, S. (2017). Solution combustion synthesis of (Co, Ni)Cr_2O_4 pigments: Influence of initial solution concentration. *Ceram. Int., 43*, 10032–10040.

González-Cortés, S. L., & Imbert, F. E. (2013). Fundamentals, properties and applications of solid catalysts prepared by Solution Combustion Synthesis (SCS). *Appl. Catal. A: Gen., 452*, 117–131.

González-Cortés, S. L., Qian, Y., Almegren, H. A., Xiao, T., Kuznetsov, V. L., & Edwards, P. P. (2015). Citric acid-assisted synthesis of γ-alumina-supported high loading CoMo sulfide catalysts for the hydrodesulfurization (HDS) and hydrodenitrogenation (HDN) reactions. *Appl. Petrochem. Res., 5*, 181–197.

González-Cortés, S. L., Rodulfo-Baechler, S. R., & Imbert, F. E. (2014). Solution combustion method: A convenient approach for preparing Ni promoted Mo and MoW sulphide hydrotreating catalysts. In J. M. Grier (Ed.), *Combustion* (pp. 269–288). Hauppauge: Nova Science Publishers, Inc.

González-Cortés, S. L., Xiao, T., Costa, P. M. F. J., Fontal, B., & Green, M. L. H. (2004). Urea–organic matrix method: An alternative approach to prepare Co-MoS_2/γ-Al_2O_3 HDS catalyst. *Appl. Catal. A: Gen., 270*, 209–222.

González-Cortés, S. L., Xiao, T., Lin, T.-W., & Green, M. L. H. (2006). Influence of double promotion on HDS catalysts prepared by urea-matrix combustion synthesis. *Appl. Catal. A: Gen., 302*, 264–273.

González-Cortés, S. L., Xiao, T., Rodulfo-Baechler, S. M. A., & Green, M. L. H. (2005). Impact of the urea–matrix combustion method on the HDS performance of Ni-MoS_2/γ-Al_2O_3 catalysts. *J. Molec. Catal. A: Chem., 240*, 214–225.

Graedel, T. E. (2002). Green chemistry and sustainable development. In J. Clark & D. J. Macquarrie (Eds.), *Handbook of Green Chemistry and Technology* (pp. 56–61). New York: Wiley.

Hagen, J. (2015). *Industrial Catalysis: A Practical Approach* (3rd ed.). Weinheim: Wiley-VCH Verlag GmbH & Co. KGaA.

Heinrichs, B., Lambert, S., Job, N., & Pirard, J.-P. (2007). Sol-gel synthesis of supported metals. In J. Regalbuto (Ed.), *Catalyst Preparation Science And Engineering* (pp. 163–208). London: CRC Press.

Huang, T., Zhao, C., Qiu, Z., Luo, J., & Hu, Z. (2017). Hierarchical porous $ZnMn_2O_4$ synthesized by the sucrose-assisted combustion method for high-rate supercapacitors. *Ionics, 23*, 139–146.

Hwang, C.-C., & Wu, T.-Y. (2004). Combustion synthesis of nanocrystalline ZnO powders using zinc nitrate and glycine as reactants — influence of reactant composition. *J. Mater. Sci., 39*, 6111–6115.

Jaimeewong, P., Promsawat, M., Jiansirisomboon, S., & Watcharapasorn, A. (2016). Influence of pH values on the surface and properties of BCZT nanopowders synthesized via sol-gel auto-combustion method. *Surf. Coat. Tech.*, *306*, 16–20.

Jain, S. R., Adiga, K. C., & Pai Verneker, V. R. (1981). A new approach to thermochemical calculations of condensed fuel-oxidizer mixtures. *Combust. Flame*, *40*, 71–79.

Jian, X. M., Wenren, H. Q., Huang, S., Shi, S. J., Wang, X. L., Gu, C. D., & Tu, J. P. (2014). Oxalic acid-assisted combustion synthesized LiVO$_3$ for lithium ion batteries cathode material. *J. Power Sour.*, *246*, 417–422.

Jimenez-Gonzalez, C., Constable, D. J. C., & Ponder, C. S. (2012). Evaluating the "Greenness" of chemical processes and products in the pharmaceutical industry — a green metrics primer. *Chem. Soc. Rev.*, *41*, 1485–1498.

Jimenez-Gonzalez, C., & Overcash, M. R. (2014). The evolution of life cycle assessment in pharmaceutical and chemical applications — a perspective. *Green Chem.*, *16*, 3392–3400.

Jimenez-Gonzalez, C., Ponder, C. S., Broxterman, Q. B., & Manley, J. B. (2011). Using the right green yardstick: Why process mass intensity is used in the pharmaceutical industry to drive more sustainable processes. *Org. Process Res. Dev.*, *15*, 912–917.

Jones, K. (1973). *The Chemistry of Nitrogen, Comprehensive Inorganic Chemistry*. Oxford: Pergamon Press.

Kang, W., Ozgur, D. O., & Varma, A. (2018). Solution combustion synthesis of high surface area CeO$_2$ nanopowders for catalytic applications: Reaction mechanism and properties. *ACS Appl. Nano Mater.*, *1*, 675–685.

Kaur, P., Chawla, S. K., Meena, S. S., Yusuf, S. M., & Bindra Narang, S. (2016). Synthesis of Co-Zr doped nanocrystalline strontium hexaferrites by sol-gel auto-combustion route using sucrose as fuel and study of their structural, magnetic and electrical properties. *Ceram. Int.*, *42*, 14475–14489.

Kaushik, M., & Moores, A. (2017). New trends in sustainable nanocatalysis: Emerging use of earth abundant metals. *Curr. Opin. Green Sustain. Chem.*, *7*, 39–45.

Khaliullin, S. M., Zhuravlev, V. D., & Bamburov, V. G. (2016). Solution-combustion synthesis of oxide nanoparticles from nitrate solutions containing glycine and urea: Thermodynamic aspects. *Int. J. Self-Propag. High-Temp. Synth.*, *25*, 139–148.

Khort, A., Podboloto, K., Serrano-García, R., & Gun'ko, Y. K. (2017). One-step solution combustion synthesis of pure Ni nanopowders with enhanced coercivity: The fuel effect. *J. Solid State Chem.*, *253*, 270–276.

Kim, H.-W., Kim, H.-E., Kim, H.-W., & Knowles, J. C. (2005). Improvement of hydroxyapatite sol–gel coating on titanium with ammonium hydroxide addition. *J. Am. Ceram. Soc.*, *88*, 154–159.

King, D. A. (2016). Biggest opportunity of our age. *Science*, *351*, 109.

Kingsley, J. J., & Patil, K. C. (1988). A novel combustion process for the synthesis of fine particle α-alumina and related oxide materials. *Mater. Lett.*, *6*, 427–432.

Kingsley, J. J., & Pederson, L. R. (1993). Combustion synthesis of perovskite LnCrO$_3$ powders using ammonium dichromate. *Mater. Lett.*, *18*, 89–96.

Komova, O. V., Simagina, V. I., Mukha, S. A., Netskina, O. V., Odegova, G. V., Bulavchenko, O. A., . . . Pochtar, A. A. (2016). A modified glycine–nitrate combustion method for one-step synthesis of LaFeO$_3$. *Adv. Powder Technol.*, *27*, 496–503.

Krawczuk, A., & Stadnicka, K. (2007). Hydrogen bonding in diaquatetrakis-(urea-κO)MII dinitrates, with M = Ni and Co. *Acta Crystallogr. C*, *63*, m448–m450.

Kumar, A. (2019). Current trends in cellulose assisted combustion synthesis of catalytically active nanoparticles. *Ind. Eng. Chem. Res.*, *58*, 7681–7689.

Kumar, A., Wolf, E. E., & Mukasyan, A. S. (2011). Solution combustion synthesis of metal nanopowders: Nickel-reaction pathways. *AIChE J.*, *57*, 2207–2214.

Li, C.-J., & Trost, B. M. (2008). Green chemistry for chemical synthesis. *Proc. Natl. Acad. Sci.*, *105*, 13197–13202.

Li, F., Ran, J., Jaroniec, M., & Qiao, S.-Z. (2015). Solution combustion synthesis of metal oxide nanomaterials for energy storage and conversion. *Nanoscale*, *7*, 17590–17610.

Li, H., Zhang, S., Wei, X., Yang, P., Z., J., & Meng, J. (2016). Glucose-assisted combustion synthesis of Li$_{1.2}$Ni$_{0.13}$Co$_{0.13}$Mn$_{0.54}$O$_2$ cathode materials with superior electrochemical performance for lithium-ion batteries. *RSC Adv.*, *6*, 79050–79057.

Ludwig, J. R., & Schindler, C. S. (2017). Catalyst: Sustainable catalysis. *Chem.*, *2*, 313–316.

Lwin, N., Othman, R., Noor, A. F. M., Sreekantan, S., Yong, T. C., Singh, R., & Tin, C.-C. (2015). Influence of pH on the physical and electromagnetic properties of Mg–Mn ferrite synthesized by a solution combustion method. *Mater. Charact.*, *110*, 109–115.

Malecka, B., Lacz, A., Drozdz, E., & Małecki, A. (2015). Thermal decomposition of d-metal nitrates supported on alumina. *J. Therm. Anal. Calorim.*, *119*, 1053–1061.

Mali, A., & Ataie, A. (2005). Structural characterization of nano-crystalline BaFe$_{12}$O$_{19}$ powders synthesized by sol–gel combustion route. *Scr. Mater.*, *53*, 1065–1070.

Manjunath, G., Vardhan, R. V., Salian, A., Jagannatha, R., Kedia, M., & Mandal, S. (2018). Effect of annealing-temperature-assisted phase evolution on conductivity of solution combustion processed calcium vanadium oxide films. *Bull. Mater. Sci.*, *41*, 126.

Manukyan, K. V., Cross, A., Roslyakov, S., Rouvimov, S., Rogachev, A. S., Wolf, E. E., & Mukasyan, A. S. (2013). Solution combustion synthesis of nano-crystalline metallic materials: Mechanistic studies. *J. Phys. Chem. C.*, *117*, 24417–24427.

McElroy, C. R., Constantinou, A., Jones, L. C., Summerton, L., & Clark, J. H. (2015). Towards a holistic approach to metrics for the 21st century pharmaceutical industry. *Green Chem.*, *17*, 3111–3121.

Merzhanov, A. G. (1996). Combustion processes that synthesize materials. *J. Mater. Proc. Technol.*, *56*, 222–241.

Merzhanov, A. G. (2004). The chemistry of self-propagating high-temperature synthesis. *J. Mater Chem., 14*, 1779–1786.

Metzger, J. O. (1998). Solvent-free organic syntheses. *Angew. Chem. Int. Ed., 37*, 2975–2978.

Mishra, R. R., & Sharma, A. K. (2016). Microwave–material interaction phenomena: Heating mechanisms, challenges and opportunities in material processing. *Compos. Part A Appl. Sci. Manuf., 81*, 78–97.

Moore, J. J., & Feng, H. J. (1995a). Combustion synthesis of advanced materials: Parte I. Reaction parameter. *Prog. Mater. Sci., 39*, 275–316.

Moore, J. J., & Feng, H. J. (1995b). Combustion synthesis of advanced materials: Parte II. Classification, applications and modelling. *Prog. Mater. Sci., 39*, 243–273.

Moraes, G. G., Pozzobom, I. E. F., Fernandes, C. P., & de Oliveira, A. P. N. (2015). MgAl$_2$O$_4$ foams obtained by combustion synthesis. *Chem. Eng. Trans., 43*, 1801–1806.

Morsi, K. (2012). The diversity of combustion synthesis processing: A review. *J. Mater. Sci., 7*, 68–92.

Mukasyan, A. S., & Dinka, P. (2007). Novel approaches to solution-combustion synthesis of nanomaterials. *Int. J. Self-Propag. High-Temp. Synth., 16*, 23–35.

Mukasyan, A. S., Rogachev, A. S., & Aruna, S. T. (2015). Combustion synthesis in nanostructured reactive systems. *Adv. Powder Technol., 26*, 954–976.

Munnik, P., de Jongh, P. E., & de Jong, K. P. (2015). Recent developments in the synthesis of supported catalysts. *Chem. Rev., 115*, 6687–6718.

Muresan, L. E., Cadis, A. I., Perhaita, I., Ponta, O., & Silipas, D. T. (2015). Thermal behavior of precursors for synthesis of Y$_2$SiO$_5$:Ce phosphor via gel combustion. *J. Therm. Anal. Calor., 119*, 1565–1576.

Nair, S. R., Purohit, R. D., Sinha, P. K., & Tyagi, A. K. (2009). Sr-doped LaCoO$_3$ through acetate–nitrate combustion: Effect of extra oxidant NH$_4$NO$_3$. *J. Alloys Comp., 477*, 644–647.

Ortega-San Martín, L., Vidal, K., Roldán-Pozo, B., Coello, Y., Larrañaga, A., & Arriortua, M. I. (2016). Synthesis method dependence of the lattice effects in Ln$_{0.5}$M$_{0.5}$FeO$_3$ perovskites (Ln = La and (Nd or Gd); M = Ba and (Ca or Sr)). *Mater. Res. Express, 3*, 056302.

Patil, K. C., Aruna, S. T., & Ekambaram, S. (1997). Combustion synthesis. *Curr. Opin. Solid State & Mater. Sci.*, 158–165.

Patil, K. C., Hedge, M. S., Rattan, R., & Aruna, S. T. (2008). *Chemistry of Nanocrystalline Oxide Materials: Combustion Synthesis, Properties and Applications*. Singapore: World Scientific.

Patil, K. C., Nesamani, C., & Verneker, V. R. P. (1982). Synthesis and characterization of metal hydrazine nitrate, azide and perchlorate complexes. *Synth. React. Inorg. Met.-Org. Chem., 12*, 383–395.

Pavithra, N. S., Lingaraju, K., Raghu, G. K., & Nagaraju, G. (2017). Citrus maxima(Pomelo) juice mediated eco-friendly synthesis of ZnO nanoparticles: Applications to photocatalytic, electrochemical sensor and antibacterial activities. *Spectrochim. Acta A Mol. Biomol. Spectrosc., 185*, 11–19.

Podbolotov, K. B., Khort, A. A., Tarasov, A. B., Trusov, G. V., Roslyakov, S. I., & Mukasyan, A. S. (2017). Solution combustion synthesis of copper nanopowders: The fuel effect. *Combust. Sci. Technol.*, *189*, 1878–1890.

Pourgolmohammad, B., Masoudpanah, S. M., & Aboutalebi, M. R. (2017a). Effect of starting solution acidity on the characteristics of $CoFe_2O_4$ powders prepared by solution combustion synthesis. *J. Magn. Magn. Mater.*, *424*, 352–358.

Pourgolmohammad, B., Masoudpanah, S. M., & Aboutalebi, M. R. (2017b). Synthesis of $CoFe_2O_4$ powders with high surface area by solution combustion method: Effect of fuel content and cobalt precursor. *Ceram. Int.*, *43*, 3797–3803.

Prat, D., Hayler, J., & Wells, A. (2014). A survey of solvent selection guides. *Green Chem.*, *16*, 4546–4551.

Prior, T. J., & Kift, R. L. (2009). Synthesis and crystal structures of two metal urea nitrates. *J. Chem. Crystallogr.*, *39*, 558–563.

Ringuede, A., Labrincha, J. A., & Frade, J. R. (2001). A combustion synthesis method to obtain alternative cermet materials for SOFC anodes. *Solid State Ionics*, *141–142*, 549–557.

Rodriguez-Sulbaran, P. J., Lugo, C. A., Perez, M. A., Gonzalez-Cortes, S. L., D'Angelo, R., Rondon, J., . . . Del Castillo, H. L. (2018). Dry reforming of methane on LaSrNiAl perovskite-type structures synthesized by solution combustion. In S. Gonzalez-Cortes & F. E. Imbert (Eds.), *Advanced Solid Catalysts for Renewable Energy Production* (pp. 242–266). Hershey: IGI Global.

Rodulfo-Baechler, S. M. A., Gonzalez-Cortes, S., Xiao, T., Al-Megren, H. A., & Edwards, P. P. (2018). Perspective on the deep hydrotreating of renewable and non-renewable oils. In S. Gonzalez-Cortes & F. E. Imbert (Eds.), *Advanced Solid Catalysts for Renewable Energy Production* (pp. 61–94). Hershey: IGI Global.

Roy, P., Nei, D., Orikasa, T., Xu, Q., Okadome, H., Nakamura, N., & Shiina, T. (2009). A review of life cycle assessment (LCA) on some food products. *J. Food Eng.*, *90*, 1–10.

Sekar, M. A., Dhanaraj, G., Bhat, H. L., & Patil, K. C. (1992). Synthesis of fine-particle titanates by the pyrolysis of oxalate precursors. *J. Mater. Sci: Mater. Electron.*, *3*, 237–239.

Shah, M., Degenstein, N., Zanfir, M., Solunke, R., Kumar, R., Bugayong, J., & Burgers, K. (2012, September). Near-zero emissions oxy-combustion flue gas purification. *Final Report*, 2011–2210.

Sheldon, R. A. (1992). Organic synthesis: Past, present and future. *Chem. Ind.*, 903–906.

Sheldon, R. A. (2016). Engineering a more sustainable world through catalysis and green chemistry. *J. R. Soc. Interface*, *13*, 20160087.

Sheldon, R. A. (2017). The E factor 25 years on: The rise of green chemistry and sustainability. *Green Chem.*, *19*, 18–43.

Sheldon, R. A. (2018). Metrics of green chemistry and sustainability: Past, present, and future. *ACS Sustain. Chem. Eng.*, *6*, 32–48.

Sheldon, R. A., Arends, I., & Hanefeld, U. (2007). *Green Chemistry and Catalysis*. Weinheim: Wiley-VCH Verlag GmbH & Co. KGaA.

Shirvani, T., Yan, X., Inderwildi, O. R., Edwards, P. P., & King, D. A. (2011). Life cycle energy and greenhouse gas analysis for algae-derived biodiesel. *Energy Environ. Sci.*, *4*, 3773–3778.

Silas, K., Ghani, W. A. W. A. K., Choong, T. S. Y., & Rashid, U. (2019). Carbonaceous materials modified catalysts for simultaneous SO_2/NOx removal from flue gas: A review. *Catal. Rev.*, *61*, 134–161.

Specchia, S., Finocchio, E., Busca, G., & Specchia, V. (2010). Combustion synthesis. In M. Lackner, F. Winter, & A. K. Agarwal (Eds.), *Handbook of Combustion* (pp. 439–472). Weinheim: Wiley-VCH Verlag GmbH & Co. KGaA.

Srikesh, G., & Nesaraj, A. S. (2015). Synthesis and characterization of phase pure NiO nanoparticles via the combustion route using different organic fuels for electrochemical capacitor applications. *J. Electrochem. Sci. Technol.*, *6*, 16–25.

Stuerga, D. (2006). Microwave–material Interactions and dielectric properties, key ingredients for mastery of chemical microwave processes. In A. Loupy (Ed.), *Microwaves in Organic Synthesis* (2nd ed., pp. 1–61). Weinheim: Wiley-VCH Verlag GmbH & Co. KGaA.

Sutka, A., & Mezinskis, G. (2012). Sol-gel auto-combustion synthesis of spinel-type ferrite nanomaterials. *Front. Mater. Sci.*, *6*, 128–141.

Thoda, O., Xanthopoulou, G., Vekinis, G., & Chroneos, A. (2018). Review of recent studies on solution combustion synthesis of nanostructured catalysts. *Adv. Eng. Mater.*, *20*, 1800047.

Thoda, O., Xanthopoulou, G., Vekinis, G., & Chroneos, A. (2019). The effect of the precursor solution's pretreatment on the properties and microstructure of the SCS final nanomaterials. *Appl. Sci.*, *9*, 1200.

Trost, B. M. (1991). The atom economy: A search for synthetic efficiency. *Science*, *254*, 1471–1477.

Trusov, G. V., Tarasov, A. B., Goodilin, E. A., Rogachev, A. S., Roslyakov, S. I., Rouvimov, S., ... Mukasyan, A. S. (2016). Spray solution combustion synthesis of metallic hollow microspheres. *J. Phys. Chem. C*, *120*, 7165–7171.

Uneyama, H., Kobayashi, H., & Tonouchi, N. (2017). New functions and potential applications of amino acids. In M. I. A. Yokota (Ed.), *Amino Acid Fermentation*. Tokyo: Springer.

Varma, A., Mukasyan, A. S., Rogachev, A. S., & Manukyan, K. V. (2016). Solution combustion synthesis of nanoscale materials. *Chem. Rev.*, *116*, 14493–14586.

Voskanyan, A. A., Chan, K.-Y., & Li, C.-Y. V. (2016). Colloidal solution combustion synthesis: Toward mass production of a crystalline Uniform mesoporous CeO_2 catalyst with tunable porosity. *Chem. Mater.*, *28*, 2768–2775.

Wolf, E. E., Kumar, A., & Mukasyan, A. S. (2019). Combustion synthesis: A novel method of catalyst preparation. *Catalysis*, *31*, 297–346.

Wu, K. H., Yu, C. H., Chang, Y. C., & Horng, D. N. (2004). Effect of pH on the formation and combustion process of sol–gel auto-combustion derived Ni Zn ferrite/SiO$_2$ composites. *J. Solid State Chem., 177*, 4119–4125.

Xanthopoulou, G., Marinou, A., Karanasios, K., & Vekinis, G. (2017a). Combustion synthesis during flame spraying ("CAFSY") for the production of catalysts on substrates. *Coatings, 7*, 14.

Xanthopoulou, G., Thoda, O., Metaxa, E. D., Vekinis, G., & Chroneos, A. (2017b). Influence of atomic structure on the nano-nickel-based catalyst activity produced by solution combustion synthesis in the hydrogenation of maleic acid. *J. Catal., 348*, 9–21.

Xanthopoulou, G., Thoda, O., Roslyakov, S., Steinman, A., Kovalev, D., Levashov, E., ... Chroneos, A. (2018). Solution combustion synthesis of nano-catalysts with a hierarchical structure. *J. Catal., 364*, 112–124.

Yu, X., Smith, J., Zhou, N., Zeng, L., Guo, P., Xia, Y., ... Facchetti, A. (2015). Spray-combustion synthesis: Efficient solution route to high-performance oxide transistors. *Proc. Natl. Acad. Sci., 112*, 3217–3222.

Yuvaraj, S., Fan-Yuan, L., Tsong-Huei, C., & Chuin-Tih, Y. (2003). Thermal decomposition of metal nitrates in air and hydrogen environments. *J. Phys. Chem. B, 107*, 1044–1047.

Zavyalova, U., Scholz, P., & Ondruschka, B. (2007). Influence of cobalt precursor and fuels on the performance of combustion synthesized Co$_3$O$_4$/γ-Al$_2$O$_3$ catalysts for total oxidation of methane. *Appl. Catal. A: Gen., 323*, 226–233.

Zhao, Y., Qin, X., Yao, B., Gonzalez-Cortes, S., & Xiao, T. (2019). Tailoring the crystallite size of Co$_3$O$_4$/SiO$_2$ catalyst using organic-metal matrix method. *Catal Today*, https://doi.org/10.1016/j.cattod.2018.06.050.

Zhou, W., Shao, Z., Ran, R., Gu, H., Jin, W., & Xu, N. (2008). LSCF nanopowder from cellulose–glycine–nitrate process and its application in intermediate-temperature solid oxide fuel cells. *J. Am. Ceram. Soc., 91*, 1155–1162.

Zhou, W., Shao, Z. P., Ran, R., Jin, W. Q., & Xu, N. P. (2008). Functional nano-composite oxides synthesized by environmental-friendly auto-combustion within a micro-bioreactor. *Mater. Res. Bull., 43*, 2248–2259.

Chapter 2

Microwave-Assisted Solution Combustion Synthesis of Nanostructured Catalysts

Deshetti Jampaiah*, Perala Venkataswamy[†], and Benjaram M. Reddy[‡]

*Catalysis and Fine Chemicals Department, CSIR-Indian Institute of
Chemical Technology, Uppal Road, Hyderabad 500007, India*
*sampathdeshetti@gmail.com
[†]pvschemou07@gmail.com
[‡]bmreddy@iict.res.in

2.1. Introduction

Solution combustion synthesis (SCS) is a simple, efficient, and versatile chemical method for fabricating nanostructured metal oxides with controllable properties for a wide range of catalytic applications (Varma *et al.*, 2016; Manukyan *et al.*, 2017; Asefi *et al.*, 2018; Kalantari *et al.*, 2018; Kombaiah *et al.*, 2019, Specchia *et al.*, 2017; Mirzaei *et al.*, 2016). Compared with other solution-based methods (sol-gel, solvothermal, hydrothermal, etc.), the SCS method offers several advantages such as saving of energy and time, minimizing waste production, and desired quality of the product (Patil *et al.*, 2002; Rogachev *et al.*, 2007). The SCS method involves a self-sustained exothermic redox reaction between different oxidizers (metal nitrate precursors) and additive fuels (glycine, citric acid, urea, ethylene glycol, sorbitol, hydrazine, etc.). The general process of the SCS method is shown in Fig. 2.1. In a typical SCS method, the reaction between fuel- and oxygen-containing species, formed during the decomposition of the nitrate species, provides high temperature and rapid interaction, where the fuel can form

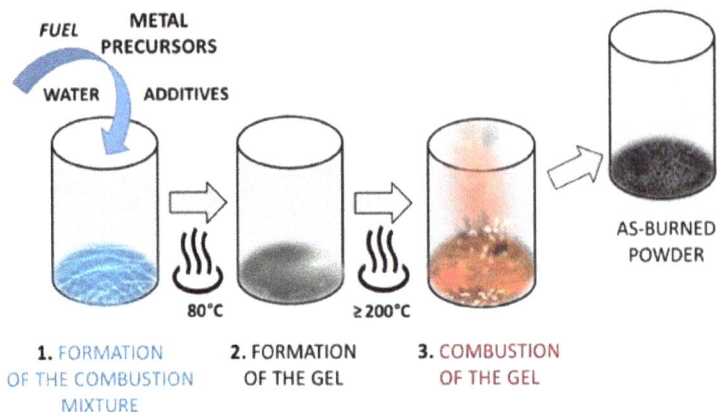

Figure 2.1.　Three main steps in the SCS. (Reprinted with permission from Deganello and Tyagi (2018). Copyright © Elsevier.)

complexes homogeneously with the metal nitrate precursors. The homogeneous mixing of precursors in the liquid state can lead to the desired stoichiometry of the products. The combustion process produces various gases from this homogeneous mixture, thereby generating heat for the desired reaction. The heat generated from the fuel is sufficient for obtaining nanostructured metal oxides. This implies that no postprocess treatment like calcination is required to obtain the materials. Although there is no need of calcination process during combustion, there might be a chance of secondary phases and/or partially combusted carbon and/or partially decomposed organics formation on the catalyst surface because of the usage of excessive fuel (Deganello and Tyagi, 2018). Consequently, the carbon formation reduces the specific surface area of the catalyst. However, it can be controlled by changing the amount of fuel to metal nitrate ratios, which enables to produce the desired nanostructures with high specific surface area, morphology, and crystallography. Table 2.1 summarizes some examples of oxidizer, fuel, and solvent used in the SCS methods.

There are several heating or ignition sources such as furnace, ultrasound, oven, and microwave that can be used as an external energy to facilitate the SCS reaction. Among them, microwave heating is the best available energy to preheat the precursor solutions

Table 2.1. Frequently used components involved in the SCS method.

Solvent	Oxidizer	Fuel
Water, benzene, methanol, ethanol, formaldehyde, and so on	Metal nitrates, metal nitrate hydrates, ammonium metal nitrates, and nitric acid	Urea, glycine, citric acid, acetonitrile, acetyl acetone, glucose, sucrose, and so on

to self-ignition temperature. The microwave treatment can generate heat because of the interaction of electromagnetic field with the electric and magnetic dipoles. High temperature will be supplied to the metal precursors because of uniform interaction of microwave with the precursor materials, which lead to facile reaction. These conditions favor shorter reaction times, reduction in energy, and high product yields.

Nanostructured materials have attracted a great interest because of their unique characteristics different from bulk materials. The choice of the SCS method is crucial for predicting the properties and their implementation in different applications. A wide range of synthetic methods are available in the literature with regard to the preparation of nanostructured materials, including high-energy milling, sol-gel, hydrothermal, copolymer synthesis, and freeze drying (Baláž *et al.*, 2004, Cheng *et al.*, 2004, Chen *et al.*, 2004, Choi *et al.*, 2003, Vie *et al.*, 2004). The major problems associated with these techniques are either the process duration or the difficulties in achieving the desired phase composition of the product.

Alternatively, microwave heating method for fabricating nanostructured materials has been increasingly investigated because of its advantage over traditional methods (McBride *et al.*, 1994; Rosa *et al.*, 2013; Mahmoodi *et al.*, 2014). The microwave heating technique has been regarded a "Fast Chemistry" method. Furthermore, considering the low energy requirements, the technique is also consistent with most of the "Green Chemistry" principles: use of non-hazardous reactants and solvents, high efficiency in terms of energy consumption *vs.* yield, and easy monitoring to prevent pollution (Anastas, 2007).

Unlike conventional heating, the following are the major advantages associated with the microwave heating for chemical synthesis:

- Rapid energy transfer.
- Uniformity of heating.
- Reduced power consumption.
- Quick start-up and stopping.
- Short reaction time that results in lower operating and processing costs.
- No contact between heating source and reacting system (reagents or solvents).
- Excellent control of reaction parameters.
- Microwaves can penetrate materials and deposit energy; heat can be transmitted throughout the material.
- In multiple-phase materials, some phases may couple more readily with the microwaves. Thus, processing materials with new microstructures by selectively heating the distinct phases may be possible.
- Microwaves can also initiate chemical reactions not possible in conventional process through selective heating of reactants. Thus, new materials may be created.
- Clean environment at the point of use.
- Cost-effective and simple procedure.

Apart from the microwave-SCS approach, there is also a possibility of catalyst synthesis using microwave energy in the absence of solution combustion. For example, $Ni_3(BO_3)_2$ nanoflowers (Ede *et al.*, 2018), carbon nanodots (Liu *et al.*, 2015), Pt@CeO$_2$ multicore@shell (Wang *et al.*, 2013), shuttle-shaped CeO$_2$ (Meher *et al.*, 2012), and others were prepared using microwave-assisted method without solution combustion. However, this chapter aims to provide an overview on the most significant procedures developed in the microwave-SCS field of research. It will highlight the basic principles involved in the field of the microwave-assisted SCS method. Synthesis of various types of nanostructured metal oxides through microwave-assisted method will be discussed. The potential applications of the as-prepared nanostructured catalysts will be discussed while explaining the methodology in this chapter.

2.2. Microwave Characteristics

Microwave heating has been known since the early 1940s and has been successfully used in the food industry. Microwave chemistry has received great attention in recent years, as its use has been started in the preparative chemistry and materials synthesis since 1986 (Gedye *et al.*, 1986; Giguere *et al.*, 1986). Use of microwave energy in green synthesis of nanomaterials has several merits and is prominent from both scientific and engineering perspectives.

2.2.1. *Fundamentals of Microwave Irradiation*

Microwave irradiation (microwave) consists of electromagnetic wave in the range between 300 MHz and 300 GHz, which corresponds to wavelengths of 100 cm to 0.1 cm. This broad frequency range leads microwave radiation to a wide variety of applications. The microwave region of the electromagnetic spectrum lies between infrared and radio frequencies (Baig and Varma, 2012; van Eldik and Hubbard, 1996). The frequency and wavelength ranges of different electromagnetic waves including microwave region are shown in Fig. 2.2.

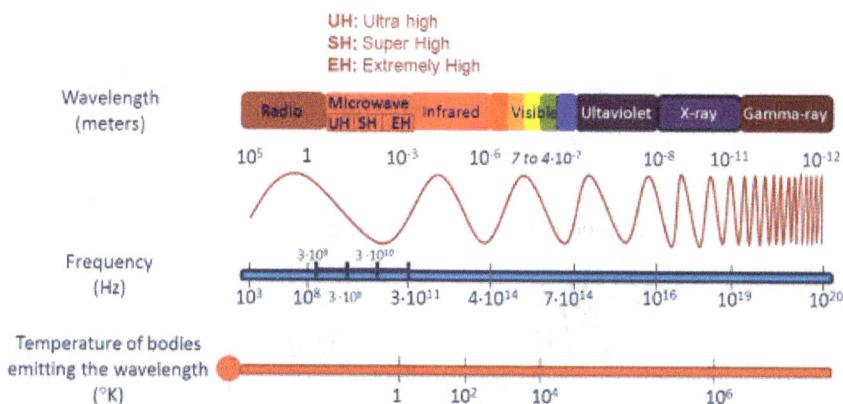

Figure 2.2. The frequency and wavelength range of electromagnetic waves. (Reprinted with permission from El Khaled *et al.* (2018). Copyright © Elsevier.)

Although there is no formal definition of the frequency range for "microwaves," some textbooks define all frequencies above 300 MHz as microwaves. The term "microwave" seems to have first appeared in 1932 by Nello Carrara in the first issue of *Alta Frequenza*. The Italian word is *microonde*. The term gained acceptance during the Second World War to describe wavelengths less than about 30 cm. All domestic microwave ovens, microwave reactors, and other laboratory and industrial systems usually work at 2.45 GHz (the corresponding wavelength is 12.24 cm), because this frequency is approved worldwide and is used in currently available commercial microwave chemistry equipment. The microwave photon corresponding to 2.45 GHz has the energy close to 0.0016 eV ($\sim$1 kJ/mol).

2.2.2. *Principles of Microwave Heating*

The interaction of dielectric materials (liquids or solids) with the microwaves has given rise to what is generally known as dielectric heating. Microwave chemistry is based on the efficient heating of materials using dielectric heating effect. Dielectric heating works through two major mechanisms, namely dipolar polarization and ionic conduction, which are shown in Fig. 2.3 (Jacob *et al.*, 1995).

- **Dipolar polarization**: For a substance to be able to generate heat when irradiated with microwaves, it must be a dipole, that

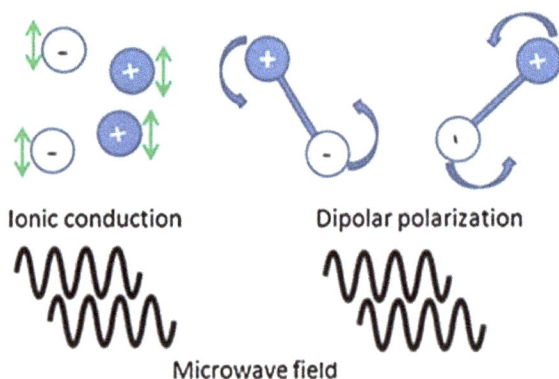

Ionic conduction   Dipolar polarization

Microwave field

Figure 2.3.   Ionic conduction and dipolar polarization under microwave conditions. (Reprinted with permission from Gude *et al.* (2013). Copyright © Springer.)

is, its molecular structure must be partly negatively and partly positively charged. As the microwave field is oscillating, the dipoles in the field align to the oscillating field. This alignment causes rotation, which results in friction and ultimately the generation of heat energy.

• **Ionic conduction**: As free ions or ionic species in the reaction solution being heated through ionic conduction mechanism, the electric field leads to ionic motion as the molecules orient themselves to the changing electric field of the microwaves.

Based on the above two mechanisms, microwaves transfer energy in 10^{-9} s (with a frequency of 10^9 Hz) with each cycle of electromagnetic energy. The kinetic molecular relaxation from this energy is approximately 10^{-5} s. Therefore, the energy transfers faster than the molecules can relax, resulting in the non-equilibrium condition and high instantaneous temperatures. As the high instantaneous heating is generated by microwave irradiation, energetic collisions are generated much faster than by conventional heating, thereby reaction rates can be increased (Rao *et al.*, 1999). This rate change can be traced to the Arrhenius equation. Arrhenius reaction rate equation is given as follows:

$$k = A \cdot e^{-E/RT} \qquad (2.1)$$

The prefactor A in the equation is in units per second, making it dependent on the frequency of the atomic vibrations at the reaction interface. The frequency of the microwaves directly alters A, which has a multiplicative effect on the reaction rate coefficient k. E is the activation energy for the reaction, which is then divided by the constant R and temperature T.

2.2.3. *Interaction of Microwave with Materials*

In general, materials fall into three categories with respect to their interaction with microwave fields (El Khaled *et al.*, 2018; Gude *et al.*, 2013). The interaction of microwaves with these materials is shown in Fig. 2.4.

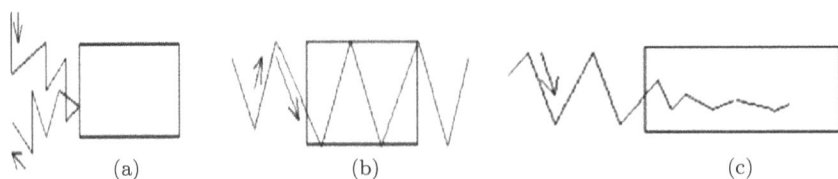

Figure 2.4. Microwave interaction with different materials: (a) reflectors, (b) transmitters, and (c) absorbers.

- Microwave reflectors, typified by bulk metals and alloys, such as brass, are therefore used in making microwave guides (Fig. 2.4a).
- Microwave transmitters, which are transparent to microwaves, typified by fused quartz, zircon, several glasses and ceramics (not containing any transition element), teflon, and so on, are therefore employed for making cookware and containers for carrying out chemical reactions in microwaves (Fig. 2.4b).

Microwave absorbers constitute the most important class of materials for microwave synthesis; they take up the energy from the microwave field and get heated up very rapidly (Fig. 2.4c). Microwave heating normally can be carried out through the interaction of electromagnetic radiation with the dipole moment of the molecules, whereas dipole moments rotate to align themselves with the alternating electric field of the microwave (Purohit *et al.*, 2001; Chang *et al.*, 1994). The ratio of the dielectric loss (loss factor) to the dielectric constant, dissipation factor (frequently called the loss tangent, tan δ), is applied to envisage a material's behavior in a microwave field. The microwave absorption ability of a material is directly proportional to its dissipation factor (Li and Yang, 2008). The microwave penetration depth, D, that is, the distance at which the power of the incident electromagnetic waves is reduced by one half, is also dependent on such factor.

2.2.4. *Components of Microwave Systems*

Design of microwave apparatus depends strongly on its purpose (Kitchen *et al.*, 2014). A microwave oven, or a microwave system, is an appliance that heats the material by dielectric heating. This

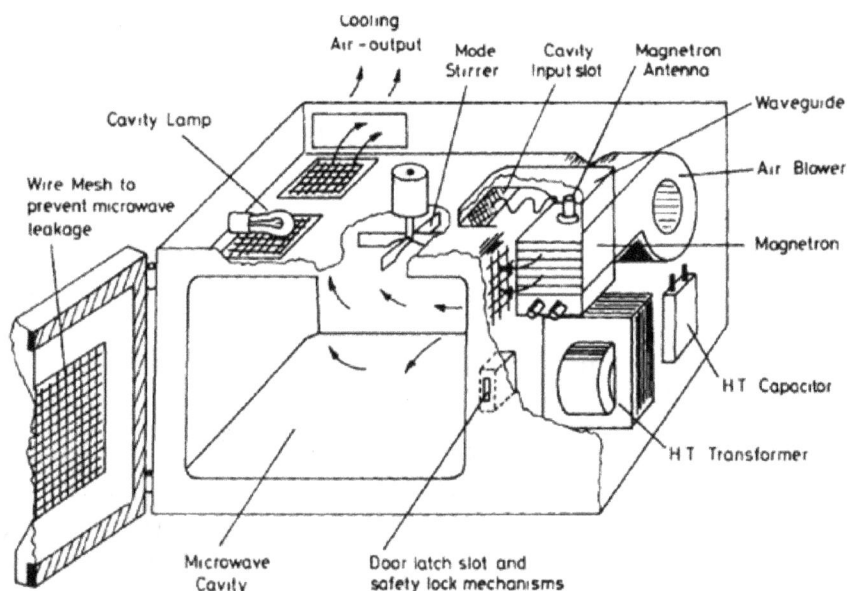

Figure 2.5. Block diagram of the modern microwave system. (Reprinted with permission from Mingos *et al.* (1991). Copyright © RSC.)

is accomplished using microwave radiation to heat water and other polarized molecules within the material.

A typical microwave system consists of:

- Source of microwave (magnetron)
- Wave guide
- Cavity (applicator), a chamber in which microwaves interact with chemical system and a system to control heating

To achieve the high power and frequencies required for microwave heating, most microwave sources are vacuum tubes. Some vacuum tubes that are used for microwave heating are magnetrons, travelling wave tubes, and klystrons. Magnetron tubes, which are used in home microwave ovens, are efficient and reliable (Purohit *et al.*, 2001; Kenji, *et al.*, 1986; Gillon *et al.*, 1987). The heart of the microwave oven is magnetron, an oscillator that converts high-voltage pulse into a pulse of microwave power (Fini and Breccia, 1999). The maximum declared efficiency of modern magnetrons operating at

2.45 GHz is approximately more than 70%, which becomes slightly higher at lower frequencies. The microwave enters a waveguide, whose reflective walls allow the transmission of the radiation into the cavity or applicator. The cavity is a type of a box in which microwaves interact with the chemical system. The type of applicator to be used for performing microwave depends on the properties of the material to be processed. The principal features of modern microwave system are shown in Fig. 2.5.

2.3. Literature Survey on the Microwave-Assisted SCS of Nanostructured Catalysts

A microwave-assisted SCS approach could produce a variety of nanostructured catalysts for energy and environmental applications in a sustainable manner. It has attracted a great deal of attention because of its many advantages, which are discussed in the earlier sections. It is not surprising that this synthetic approach has been used to produce several nanostructured catalysts, including pure metal oxides, binary and ternary metal oxides, perovskites, spinels, hetero-structured materials, and so on. Some of the important nanostructured catalysts synthesized using the microwave-assisted SCS method are summarized in the following sections.

2.3.1. *Synthesis of Pristine Metal Oxides*

Among the various transition metal oxides, zinc oxide (ZnO) has been considered an interesting metal oxide through the microwave-assisted SCS method because of its excellent electrical, optical, and magnetic properties (Rodnyi and Khodyuk, 2011). Synthesis of nanostructured ZnO has also attracted significant attention because of its wide range of photocatalytic applications (Ong *et al.*, 2018).

Nanocrystalline ZnO nanoplatelets were synthesized by Yüksel Köseoğlua (2014) using the microwave-assisted SCS method. Zinc nitrate as the oxidant and glycine as the fuel were used in the combustion process. Scanning electronic microscopy (SEM) was employed to determine the morphology. It was found that ZnO particles resemble nanoplates, which consist of 20- to 40-nm-sized particles and they were nicely stacked together. The obtained ZnO

Figure 2.6. SEM images of flower-like ZnO nanoparticles obtained by microwave-assisted combustion method with different urea/Zn^{2+} ratios: (a and b) 1:1, (c and d) 5:3, and (e and f) 3:1. (Reprinted with permission from Cao *et al.* (2011). Copyright © Elsevier.)

nanoplatelets exhibited 3.09 eV of band gap energy and showed better ferromagnetic behavior. Cao *et al.* (2011) reported flower-like ZnO nanostructures, and in the synthesis method, the authors used urea as a fuel for combustion process. It was found that by changing the molar ratio of urea to Zn^{2+} (1:1, 5:3, and 3:1), different morphologies could be obtained for ZnO nanoparticles (Fig. 2.6).

As discussed earlier, the amount of fuel and oxidant played a crucial role in determining the morphology for the required metal oxides. The flower-like morphology was found when 1:1 urea to Zn molar ratio was used irrespective of the microwave power (170, 340, and 680 W). Here, the urea acted as structure-directing agent for the formation of flower-like ZnO nanoparticles.

Hoffmann *et al.* (2016) used zinc diketonates and acetonitrile precursors for the fabrication of ZnO nanoparticles. The obtained ZnO nanoparticles were spherical in shape. Similar types of nanoparticles were prepared by Buha *et al.* (2007) through hydrothermal synthesis at 100°C and this synthesis method required a long period of time (2 days). Here, the microwave-assisted SCS method provided an advantage of synthesizing similar type of nanoparticles in 30 min.

Moreover, nanoparticles obtained using the microwave-assisted SCS method were used as starting materials for growing one-dimensional nanostructure. Kooti and Naghdi Sedeh (2013) synthesized one-dimensional ZnO nanoparticles by combining microwave and combustion process with the reaction parameters such as 700 W of microwave energy and zinc nitrate and glycine as the oxidant and fuel, respectively. The 1-hour calcination process (400°C, 500°C, and 600°C) produced one-dimensional nanorod morphology for the ZnO nanoparticles. The observed ZnO nanorods exhibited an excellent ethanol-sensing behavior.

Another important oxide is magnesium oxide (MgO), which was prepared using microwave-assisted SCS method and the properties of MgO were well compared with that of other conventional heating methods (Selvam *et al.*, 2013). In a typical synthesis, magnesium nitrate precursor and urea as fuel were used and the reaction was operated at a power of 750 W. The microwave combustion process generated MgO nanosheets, whereas the conventional heating method generated micro-cubes (Fig. 2.7). The difference between these two methods is the microwave energy. During microwave-assisted combustion process, the generated gaseous products such as N_2, CO_2, and H_2O bubble in the inner MgO microspheres, thereby the bubbles facilitate producing numerous channels called MgO nanosheets. Besides, in conventional heating, there is a temperature gradient between the heat source and the mass to heat, which restricts the temperature distribution homogeneously and results in MgO micro-cubes. Therefore, it can be concluded that the combination of microwave energy and combustion process is advantageous for synthesizing nanostructured solid catalysts.

2.3.2. *Synthesis of Mixed Metal Oxides*

Various mixed metal oxides such as Al_2O_3–ZrO_2, MoO_3–ZrO_2, and TiO_2–M_xO_y (M_xO_y = SiO_2, Al_2O_3, and ZrO_2) were advantageously prepared using the microwave-SCS method (Tahmasebi and Paydar, 2011; Samantaray and Mishra, 2011, Reddy *et al.*, 2009). The as-prepared mixed oxides exhibited better properties (crystallite size, surface area, pore volume, redox behavior, and band gap) while

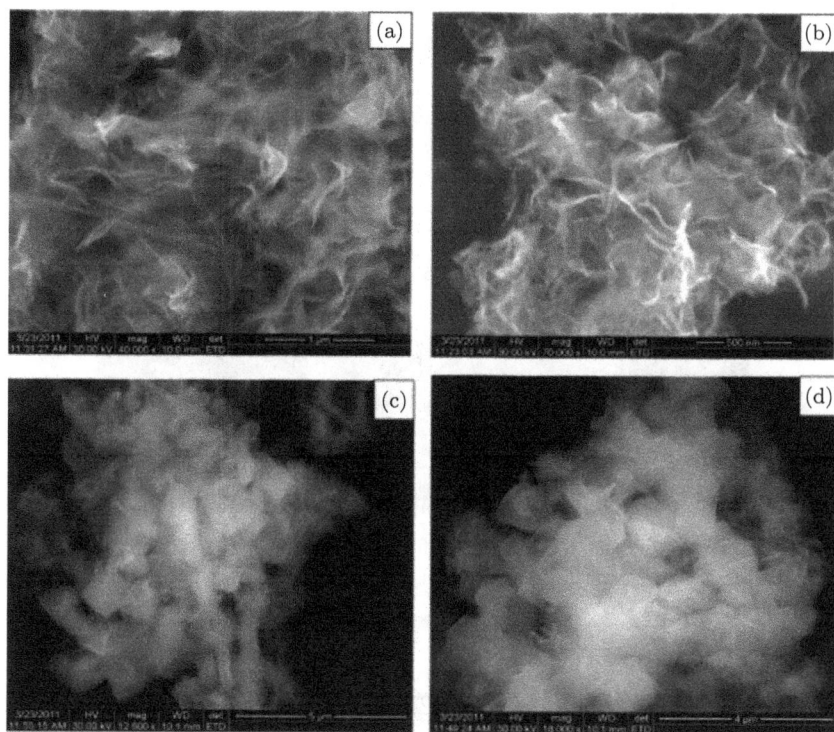

Figure 2.7. HR-SEM images of MgO: (a and b) sample A (microwave method) and (c and d) sample B (conventional method). (Reprinted with permission from Selvam *et al.* (2013). Copyright © Elsevier.)

compared to individual metal oxides. Depending on the specific catalytic application, the mixed oxides were synthesized and tested for catalytic applications. For example, ZnO has been extensively used for photocatalysis; however, the photocatalytic activity of this material was not high enough because of its larger band gap ($>3\,eV$) and the rapid recombination of charge carriers (electron and hole) during photo-excitation. Fortunately, CeO_2–ZnO was prepared by the microwave SCS method using metal precursors and urea as the fuel (Sherly *et al.*, 2015). The CeO_2–ZnO photocatalysts exhibited an excellent photocatalytic performance in the degradation of 2,4 dichlorophenol ($\sim 95\%$) in 240 min compared to pure counterparts (CeO_2 and ZnO).

Figure 2.8. Combustion reaction performed in the microwave oven during the preparation of catalysts. (Reprinted with permission from Reddy *et al.* (2012). Copyright © Elsevier.)

Reddy *et al.* (2009, 2012, 2014) and Devaiah *et al.* (2018) synthesized different ceria-based mixed oxides (CeO_2–SiO_2, CeO_2–TiO_2, CeO_2–ZrO_2, and CeO_2–Al_2O_3) using the rapid microwave-combustion approach and investigated them for CO oxidation and soot oxidation reactions. The typical microwave oven used for the combustion synthesis of CeO_2-based mixed oxides is shown in Fig. 2.8.

To prepare various CeO_2–M_xO_y (M = Si, Ti, Zr, and Al) catalysts, the authors used metal nitrates as the oxidizers and urea as the fuel during the microwave-assisted SCS approach. Required amount of the corresponding precursors (metal nitrates) were taken in a pyrex dish and dissolved in deionized water separately and mixed together. To this, the required stoichiometric quantity of urea (considered from propellant chemistry) dissolved in water was used and mixed thoroughly to obtain a homogenous solution at ambient conditions. A microwave oven used for domestic purposes was modified with an outlet for exhaust gases as shown in Fig. 2.8 (BPL, India, 2.54 GHz, BMO-700T, 700W). This modified oven was used as a heating source to initiate the combustion between metal nitrates and urea. The dish containing the metal precursors and urea was introduced into the center of microwave oven and was operated. Initially, the solution boils with slow removal of water from the system followed by decomposition and spontaneous combustion leading to flame generation for a short time. Along with the flame, there occurs liberation of various gases such as N_2, CO_2, H_2O, NH_3, and NO_2 resulting in the light-yellow solid material. The resultant powders were grinded for 10 min with mortar and pestle to obtain the homogenous powder from material.

In the SCS process, the stoichiometric amount of metal nitrate and fuel is significant for the formation of homogeneous nanoparticles. Reddy *et al.* (2009) calculated the stoichiometric amount of materials for the synthesis of different CeO_2-based mixed oxides. For example, the following equation describes the required number of precursors in moles for the preparation of CeO_2–SiO_2. For this preparation, $Ce(NO_3)_3$, $SiO(NO_3)_2$, and urea were used as the starting materials.

Ce valency in $Ce(NO_3)_3$ is $+3 + (0 - 3 \times 2) \times 3 = -15$.

Similarly, for $SiO(NO_3)_2$ and NH_2CONH_2 the valency is -10 and $+6$.

The required stoichiometric ratio of urea to nitrates is $-15 - 10/6 = 4.66$ moles as per the propellant chemistry. If we assume complete combustion, the equation can be represented as follows:

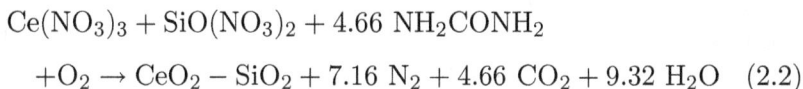

$$Ce(NO_3)_3 + SiO(NO_3)_2 + 4.66 \ NH_2CONH_2$$
$$+O_2 \rightarrow CeO_2 - SiO_2 + 7.16 \ N_2 + 4.66 \ CO_2 + 9.32 \ H_2O \quad (2.2)$$

Similarly, for the preparation of CeO_2–Al_2O_3, CeO_2–ZrO_2, and CeO_2–TiO_2-mixed oxides, the amount of urea required is 5.00, 4.66, and 4.66 moles, respectively. The respective theoretical equations are as follows:

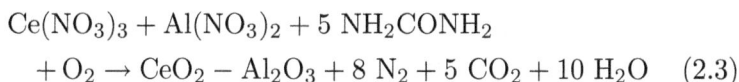

$$Ce(NO_3)_3 + Al(NO_3)_2 + 5 \ NH_2CONH_2$$
$$+ O_2 \rightarrow CeO_2 - Al_2O_3 + 8 \ N_2 + 5 \ CO_2 + 10 \ H_2O \quad (2.3)$$

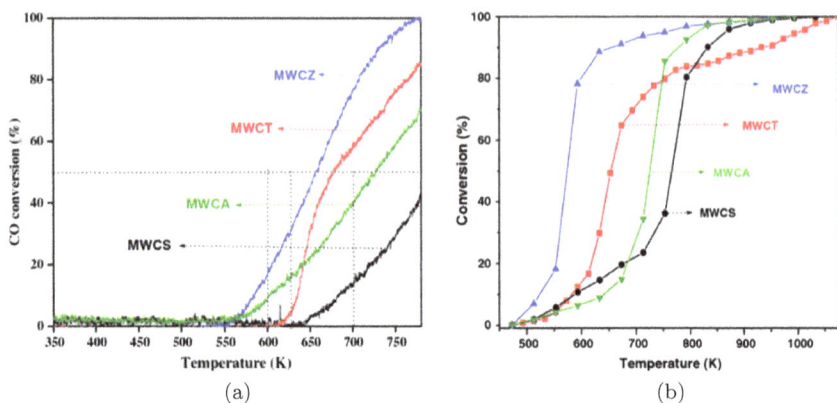

Figure 2.9. (a) CO oxidation and (b) soot oxidation activities *vs.* temperature profiles of various mixed oxides (MWCS: CeO_2–SiO_2, MWCT: CeO_2–TiO_2, MWCZ: CeO_2–ZrO_2, and MWCA: CeO_2–Al_2O_3). (Reprinted with permission from Reddy *et al.* (2009). Copyright © Springer.)

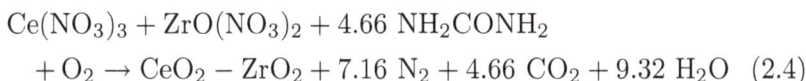

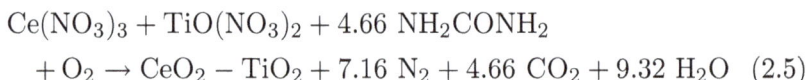

$$Ce(NO_3)_3 + ZrO(NO_3)_2 + 4.66\ NH_2CONH_2$$
$$+ O_2 \rightarrow CeO_2 - ZrO_2 + 7.16\ N_2 + 4.66\ CO_2 + 9.32\ H_2O \quad (2.4)$$

$$Ce(NO_3)_3 + TiO(NO_3)_2 + 4.66\ NH_2CONH_2$$
$$+ O_2 \rightarrow CeO_2 - TiO_2 + 7.16\ N_2 + 4.66\ CO_2 + 9.32\ H_2O \quad (2.5)$$

The as-prepared CeO_2-based mixed oxides exhibited superior catalytic performance toward CO and soot oxidation reactions. The CO and soot oxidation curves of the CeO_2–SiO_2, CeO_2–TiO_2, CeO_2–ZrO_2, and CeO_2–Al_2O_3-mixed oxide catalysts are shown in Figs. 2.9a and 2.9b, respectively (Reddy *et al.*, 2009). Among all the mixed oxides, the CeO_2–ZrO_2 showed better performance than other catalysts. The catalytic activity trend observed is as follows: CeO_2–ZrO_2 > CeO_2–TiO_2 > CeO_2–Al_2O_3 > CeO_2–SiO_2. It was concluded that the microwave-assisted synthesis has a lot of impact because of its simplicity, flexibility, and the easy control of large-scale preparation of industrial catalysts.

2.3.3. *Synthesis of Perovskites*

Perovskite oxides are defined by a general formula ABO_3, where A stands for larger cation and is surrounded by 12 oxygen atoms;

B stands for smaller cation and is surrounded by 6 oxygen atoms (Athayde *et al.*, 2016). The perovskite oxides are good alternative catalysts for noble metal-based catalysts because of their high thermal stability, lower cost, and superior catalytic activity. Various methods including hydrothermal synthesis, coprecipitation, and sol-gel were employed to prepare the perovskite oxides. However, these methods require high reaction temperatures, which could result in agglomerated nanoparticles with lower surface area. Compared to these conventional methods, the microwave-assisted SCS method could enable a good chemical homogeneity, a narrow size distribution, and high surface area for the perovskite oxide catalysts.

Esmaeilnejad-Ahranjani *et al.* (2011) prepared lanthanum manganite ($LaMnO_3$) perovskite using lanthanum nitrate precursor and sorbitol fuel and investigated for lowering the oxidation light-off temperature in automotive exhaust gas pollution control applications. The overall combustion reaction is described as follows:

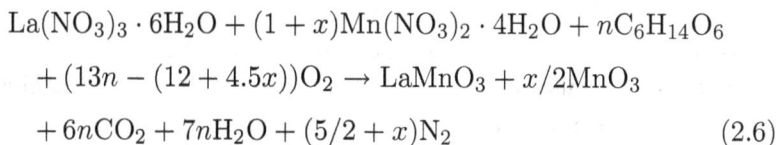

$$La(NO_3)_3 \cdot 6H_2O + (1 + x)Mn(NO_3)_2 \cdot 4H_2O + nC_6H_{14}O_6$$
$$+ (13n - (12 + 4.5x))O_2 \rightarrow LaMnO_3 + x/2MnO_3$$
$$+ 6nCO_2 + 7nH_2O + (5/2 + x)N_2 \tag{2.6}$$

SEM images of $LaMnO_3$ and $LaMn_{1.2}O_{3+\delta}$ catalysts are shown in Figs. 2.10a and 2.10b, respectively. It was found that there are large voids on the surface of nanoparticles. During the synthesis process, the microwave-induced combustion of sorbitol generates large amount of gases, which create the frothy agglomerated nanoparticles with large voids in their structure. The same research group used $LaMn_{1.2}O_{3+\delta}$ catalyst prepared using the microwave-assisted SCS method for trichloroethylene (TCE) oxidation in air (Maghsoodi *et al.*, 2013). It was found that the $LaMn_{1.2}O_{3+\delta}$ catalyst exhibits the highest catalytic activity in TCE oxidation compared with the $LaMnO_3$ catalyst. The highest amount of Mn ions inserted in the perovskite oxide enhanced the oxygen over stoichiometry, which contributed to the observed highest catalytic activity.

Other perovskite, the lithium niobate ($LiNbO_3$), also known as ferroelectric material, also has attracted wide interest in catalysis.

Figure 2.10. SEM micrographs of the calcined (a) $LaMnO_3$ and (b) $LaMn_{1.2}O_{3+\delta}$. (Reprinted with permission from Maghsoodi *et al.* (2013). Copyright © Elsevier.)

Several methods including sol-gel, mechanochemical synthesis, coprecipitation, and hydrothermal synthesis were applied to prepare $LiNbO_3$ perovskite oxide. However, these methods generate waste products to the environment along with the solid catalysts. To overcome those limitations, Carreno *et al.* (2018) used unique microwave-combustion method for fabricating $LiNbO_3$ catalyst for low-temperature aniline oxidation. The experimental diagram is shown in Fig. 2.11a. The apparatus consists of sample quartz holders, paper filters, rotors, microwave irradiation, and pressurized oxygen. The rotor designed in the apparatus has an advantage that eight samples could be combusted simultaneously. The flame temperature of the combustion reactions was measured using an optical pyrometer. This apparatus is quite different to the conventional reaction beaker setup in a microwave oven. The different preparation conditions employed to make the samples are summarized in Table 2.2.

The diagram of $LiNbO_3$ synthesis using microwave-induced combustion is shown in Fig. 2.11b. As shown, cellulose was used as the fuel along with microwave energy to prepare $LiNbO_3$ catalysts. It was expected that the combustion of cellulose provides energy for the formation of $LiNbO_3$. This catalyst prepared using microwave-induced combustion method exhibited excellent catalytic performance, and the material was prepared successfully under very short period of times (40 s to 1 min). This explains that the addition

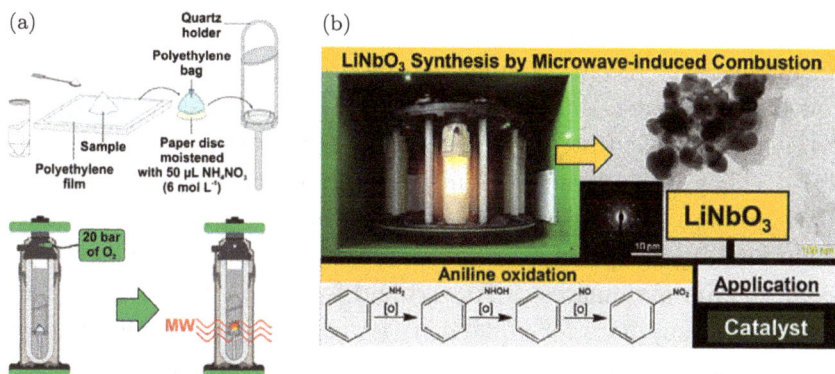

Figure 2.11. (a) Experimental procedure of the proposed minimum inhibitory concentration (MIC) method for direct LiNbO$_3$ synthesis and (b) microwave-assisted combustion synthesis of LiNbO$_3$ and the mechanism of aniline oxidation. (Reprinted with permission from Carreno *et al.* (2018). Copyright © ACS.)

Table 2.2. Preparation conditions of the samples for LiNbO$_3$ synthesis using microwave-induced combustion.

Sample	Li source	Nb source	Li:Nb ratio (mol)	Fuel	Li:Nb ratio (w/w)
1	Li$_2$CO$_3$	Nb$_2$O$_5$	1:1	Cellulose (50%) Paraffin (50%)	1:1
2	Li$_2$CO$_3$	Nb$_2$O$_5$	2:1	Cellulose (50%) Paraffin (50%)	1:1
3	Li$_2$CO$_3$	Nb$_2$O$_5$	4:1	Cellulose (50%) Paraffin (50%)	1:1
4	Li$_2$CO$_3$	Nb$_2$O$_5$	2:1	Cellulose (50%) Paraffin (50%)	1:1

Reprinted with permission from Carreno *et al.* (2018). Copyright © ACS.

of cellulose as a fuel in this method is very promising and can promote tremendous time and energy savings compared with other typical microwave-induced combustion synthesis.

2.3.4. *Synthesis of Spinel Oxides*

Spinel oxides are generally represented with the formulae M^{2+}(M^{3+})$_2$O$_4$, where M^{2+} can be Mn^{2+}, Cu^{2+}, Co^{2+}, Zn^{2+}, and so on, and M^{3+} can be Al^{3+}, Fe^{3+}, Mo^{3+}, and so on. In the formulae, M^{2+} and M^{3+} atoms can be group 2, group 13, and first-row transition metals, whereas two crystallographic sites, octahedral (O_h) and tetrahedral (T_d), are in this structure (Kharisov *et al.*, 2014). They have attracted significant attention in catalysis because of their high thermal and chemical stability. These spinel oxides are nontoxic, inexpensive, and resistant to acids and alkalis. Major examples are aluminates (MAl_2O_4) and ferrites (MFe_2O_4). They are also used as supports for noble metals to suppress the agglomeration of nanoparticles. Spinel oxides with low surface area and nonhomogeneous structure are often produced using conventional combustion methods (CCMs) at high reaction temperatures ($>400°C$) and long reaction times (>12 h). On the contrary, high surface area spinel oxides could be rapidly (<10 min) prepared using the microwave-induced SCS method.

Bououdina *et al.* (2014) prepared copper aluminate ($CuAl_2O_4$) spinel catalyst by the microwave-induced SCS method using the plant extract called aloe vera as a fuel without any template or surfactant. The structural, optical, and catalytic properties of the microwave-induced SCS method $CuAl_2O_4$ catalyst were compared with the CCM. SEM images of $CuAl_2O_4$ catalyst prepared using CCM and MWCM are shown in Fig. 2.12. It was found that the nanoparticles prepared from CCM resemble like agglomerates, whereas the MWCM-derived $CuAl_2O_4$ nanospheres clearly illustrate the significance of microwave combustion method. The determined surface area of $CuAl_2O_4$ nanospheres is $47.34\,\text{m}^2\,\text{g}^{-1}$, which is more than that of CCM sample ($15.73\,\text{m}^2\,\text{g}^{-1}$). The two catalysts were investigated for benzyl alcohol oxidation in acetonitrile medium and the corresponding conversions and selectivity are shown in Fig. 2.13. The $CuAl_2O_4$ prepared using the microwave-induced SCS method displayed more conversion and high selectivity compared with catalyst prepared using other method. Thus, it can be concluded that the presence of microwave heating enables the systematic crystal growth of nanoparticles and finally spherical nanoparticles were formed, which help to obtain finer crystallites, narrow pore size distribution, and high surface area.

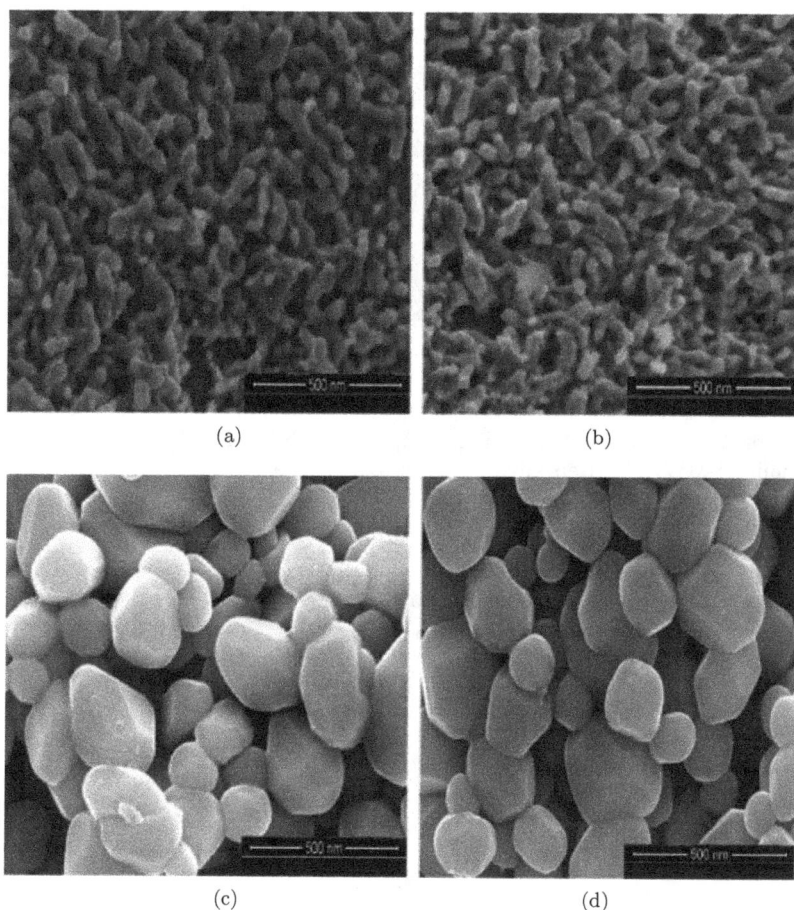

Figure 2.12. HR-SEM images of (a and b) copper aluminate-CCM and (c and d) copper aluminate-MWCM. (Reprinted with permission from Bououdina *et al.* (2014). Copyright © Elsevier.)

The other examples of aluminates include $NiAl_2O_4$ (Ragupathi *et al.*, 2014), $ZnAl_2O_4$ (Hashemzehi *et al.*, 2017), and $CoAl_2O_4$ (Zawadzki *et al.*, 2011) that were prepared using microwave method, and exhibited enhanced catalytic performance when compared with the CCM samples.

The other important spinel oxides are the ferrites, MFe_2O_4, where M can be Co^{2+}, Zn^{2+}, Cu^{2+}, Mn^{2+}, and Ni^{2+}. Manikandan *et al.* (2014) prepared cobalt ferrite ($CoFe_2O_4$) nanostructures by

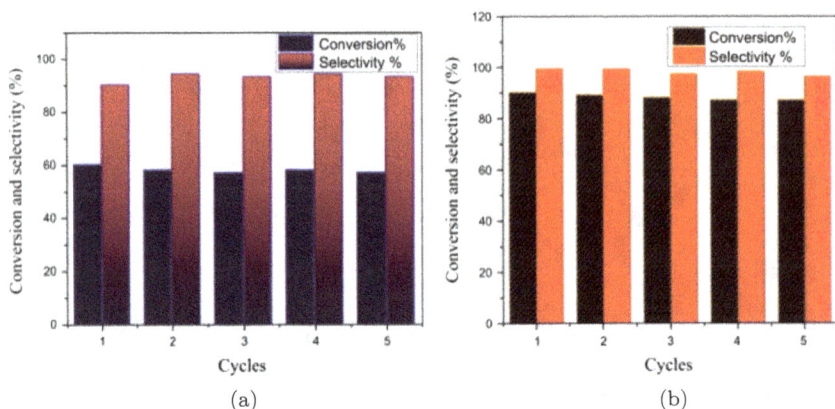

Figure 2.13. Reusability studies of (a) copper aluminate-CCM and (b) copper aluminate-MWCM. (Reprinted with permission from Bououdina *et al.* (2014). Copyright © Elsevier.)

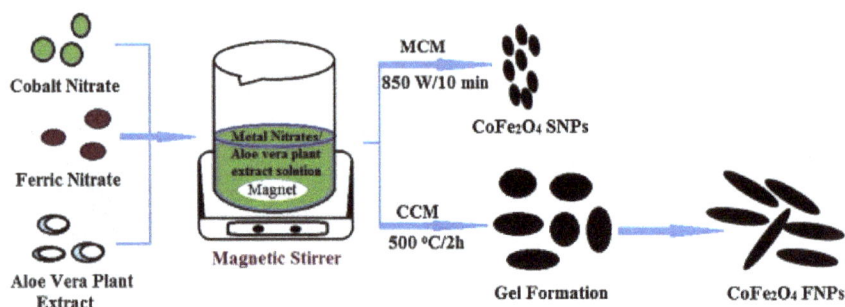

Figure 2.14. Formation mechanism of $CoFe_2O_4$ nanostructures using aloe vera plant–extracted microwave and conventional combustion synthesis. (Reprinted with permission from Manikandan *et al.* (2014). Copyright © Elsevier.)

microwave combustion method using ferric nitrate, cobalt nitrate, and aloe vera plant extract solution. As shown in Fig. 2.14, the metal nitrate precursors and plant extracts were added in a beaker and kept for vigorous stirring. After that, the reaction was carried out in a domestic microwave oven at 850 W for 10 min. Besides, the similar precursors were taken in a silica crucible and placed in an air furnace and sintered at 500°C for 2 h. The two methods resulted in different morphology for the $CoFe_2O_4$ nanoparticles. The microwave-induced SCS method produced spherical nanoparticles,

Figure 2.15. HR-SEM images of (a and b) $CoFe_2O_4$-SNPs (MCM) and (c and d) $CoFe_2O_4$-FNPs (CCM). (Reprinted with permission from Manikandan *et al.* (2014). Copyright © Elsevier.)

whereas the CCM generated flake-like nanoplatelet structures. The observed morphologies of these nanoparticles are shown in Fig. 2.15.

It can be expected that the microwave energy enables the nucleation growth of metallic Co^{2+} and Fe^{3+} cations mixture in a short period of time (10 min). This short-term treatment may avoid the agglomeration of newly formed metal nuclei, thereby the resultant nuclei were oxidized quickly to form spherical $CoFe_2O_4$ nanoparticles. As shown in Fig. 2.15, the resultant spherical nanoparticles prepared using microwave method are homogeneous and well dispersed. In conventional furnace heating method, the longer reaction time and high temperature lead to the formation of platelet structures. The

N_2 sorption experiments also proved that the microwave-induced SCS method generated $CoFe_2O_4$ nanoparticles with $93.45\,m^2\,g^{-1}$ of surface area, which was higher than that of CCM ($72.13\,m^2\,g^{-1}$). This is because the microwave irradiation heating is volumetric and homogeneous compared with conventional furnace heating, which is nonuniform and radiative. The two catalysts were tested for benzyl alcohol oxidation and found that the spherical $CoFe_2O_4$ nanoparticles exhibit 93.2% conversion with 99.5% selectivity compared with flake-like nanoplatelets (71.5% conversion and 93.4% selectivity).

The other important spinel oxides, namely $Co_{1-x}Zn_xFe_2O_4$, $CuFe_2O_4$, $Mn_{1-x}Ni_xFe_2O_4$, $Ni_xZn_{1-x}Fe_2O_4$, and $Cu_{0.5}Zn_{0.5}Fe_2O_4$ were also prepared using microwave combustion method and found that these catalysts exhibit enhanced catalytic performance in photocatalysis and antibacterial activity (Sundararajan *et al.*, 2017; Liu and Fu, 2010; Jesudoss *et al.*, 2016; Padmapriya *et al.*, 2016; Mahmoud *et al.*, 2016).

2.3.5. *Synthesis of Hetero-Composite Metal Oxides*

Heterostructures can be defined as a combination of two or more different active metal oxides, in which one metal oxide is used as a support for the other metal oxides. These heterostructures have received great attention because of their synergistic microstructural features with distinct functionalities. It is believed that the hetero structures show enhanced catalytic performance over individual metal oxides because of the interesting physicochemical and unique interfacial properties. However, obtaining homogeneous and controlled heterostructures is challenging with the existing CCMs. This section describes the importance of the microwave-induced SCS method and its advantage for the preparation of hetero-composite metal oxides as well as the effect of different fuels on the structural properties of composite metal oxide.

Ajamein *et al.* (2017) reported $CuO/ZnO/Al_2O_3$ nanocatalysts by microwave solution combustion method using metal nitrate precursors, ammonium nitrate promoter, and different fuels (sorbitol, propylene glycol, glycerol, diethylene glycol, and ethylene glycol). The synthesized $CuO/ZnO/Al_2O_3$ catalysts were tested for methanol conversion to hydrogen production. Table 2.3 lists the surface areas and catalytic activities of $CuO/ZnO/Al_2O_3$ catalysts prepared using different methods. It was found that the $CuO/ZnO/Al_2O_3$ nanocatalyst prepared by the use of ethylene glycol exhibited higher surface area and more methanol conversion. These results confirm that the employed combustion fuel in the presence of microwave energy can affect the textural properties, thereby catalytic performance.

Da Silveira *et al.* (2016) prepared $NiO-Ce_{0.75}Zr_{0.25}O_2$ nanocrystalline composite employing the microwave combustion and Pechini methods and investigated the effect of two different synthetic routes over structural and morphological properties. The Pechini method is a low-temperature synthesis route and it works on a solution polymerization. The microwave combustion and Pechini methods produced the $NiO-Ce_{0.75}Zr_{0.25}O_2$ nanocrystalline composites with 48 and $6\,m^2\,g^{-1}$ surface area, respectively. The authors tested the effect of different calcination temperatures on the surface area of the nanocrystalline composites. It was found that the $NiO-Ce_{0.75}Zr_{0.25}O_2$ prepared using microwave combustion method exhibited much more stability keeping the same surface area of $6\,m^2\,g^{-1}$ after calcination at $900°C$, whereas the Pechini method-derived catalyst provoked a significant drop from 48 to $18\,m^2\,g^{-1}$. It was concluded that the main advantages of microwave-induced combustion method compared to Pechini method are its simplicity and low cost, as well as maintaining the same structural and surface properties even after high-temperature calcination.

Rezaee and Haghighi (2016) prepared $Ce_{0.8}Zr_{0.2}O_2$ nanocatalysts over Al_2O_3 support and Reddy *et al.* (2012) prepared CuO-promoted $CeO_2-M_xO_y$ (M = Zr, La, Pr, and Sm) catalysts using microwave combustion method and tested them for CO oxidation. Other important examples of hetero structures prepared using microwave combustion method were $ZnAl_2O_4/ZnO$ composite, $KOH/Ca_{12}Al_{14}O_{33}$, $La_2NiO_4/\alpha-Al_2O_3$, and $CoAlO_x/CeO_2$ (Shahmirzaee *et al.*, 2017; Nayebzadeh *et al.*, 2017; Barros *et al.*, 2010; Fan *et al.*, 2018).

Table 2.3. Comparison of specific surface area and catalytic performance of $CuO/ZnO/Al_2O_3$ nanocatalysts in the steam methanol-reforming process.

Sample	Synthesis method	Composition	S_{BET} $(m^2\,g^{-1})$	$T(°C)$	Methanol conversion
CuO/ZnO/ Al₂O₃	Urea–nitrate combustion	38/38/24 (wt%)	42.4	260	100
CuO/ZnO/ Al₂O₃	Microwave combustion (ethylene glycol)	30/60/10 (wt%)	67.3	240	100
CuO/ZnO/ Al₂O₃	Sonochemical urea–nitrate combustion	33/33/33 (wt%)	71.8	240	100
CuO/ZnO/ Al₂O₃	Microwave combustion (sorbitol)	30/60/10 (wt%)	63.4	280	100
CuO/ZnO/ Al₂O₃	Hybrid sono-chemical –nitrate combustion	45/40/15 (wt%)	35.3	280	100
CuO/ZnO/ Al₂O₃	Microwave combustion (ethylene glycol)	30/60/10 (wt.%)	58.1	260	100
CuO/ZnO/ Al₂O₃	Microwave combustion (diethylene glycol)	30/60/10 (wt%)	56.8	280	100
CuO/ZnO/ Al₂O₃	Microwave combustion (ethylene glycol + 5% ammonium nitrate)	30/60/10 (wt%)	78.4	240	100
CuO/ZnO/ Al₂O₃	Microwave combustion (diethylene glycol + 5% ammonium nitrate)	30/60/10 (wt%)	75.3	240	100

2.3.6. *Synthesis of Other Catalysts*

Graphene is a two-dimensional material and it has been extensively studied in various energy and environmental applications because of its high chemical stability, unique electric properties, and more surface area. However, because of its high van der Waals interaction, graphene can restack together and can decrease its surface area. Synthesis of porous graphene can overcome this disadvantage and maintain its unique properties. Several methods, such as chemical activation, laser irradiation, photocatalytic oxidation, and so on, were employed to prepare the graphene with high surface area. However, these methods hardly controlled the pore size regularly and required relatively longer time with tedious steps. Wan *et al.* (2018) prepared porous graphene using the microwave-assisted SCS method to overcome these disadvantages. They reported a fast and facile way to produce porous graphene using active silver (Ag) nanoparticles under microwave irradiation. They reported that the treatment of silver acetate (CH_3COOAg) under microwave irradiation for 2 s delivered even distribution of the Ag nanoparticles on the porous graphene. These results confirm that employing microwave could deliver porous uniform nanoparticles in a very short time and this method is simple, which can open new opportunities for the synthesis of porous nanomaterials.

Mohammadzadeh *et al.* (2015) prepared Ag/ZnO photocatalyst using zinc nitrate ions as oxidizers and the mixture of glycine and glucose as the fuel under microwave treatment. The microwave-assisted SCS method produced spongy-like architectures with high porosity. Additionally, the Ag nanoparticles were decorated uniformly on the surface of ZnO. Similar photocatalysts were also prepared by Sahu *et al.* (2012) and found that this method produces uniform dispersion of Ag nanoparticles.

Yang *et al.* (2015) prepared Ag/CeO_2 catalyst using microwave-assisted biosynthesis method using *Cinnamomum camphora* leaf extract and tested this catalyst for CO oxidation. Figure 2.16 shows the process of synthesis method that produced Ag/CeO_2 catalyst. The advantage of this method is that the rapid and uniform microwave

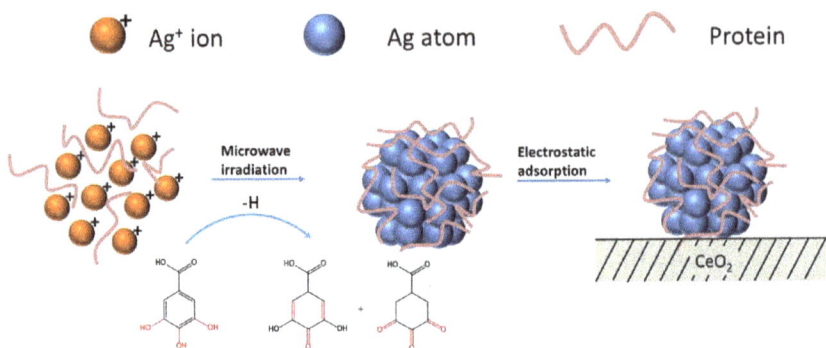

Figure 2.16. Diagram of Ag/CeO$_2$ catalyst synthesis using microwave-assisted combustion. (Reprinted with permission from Yang *et al.* (2015). Copyright ©️ Elsevier.)

heating process can generate homogeneous metal nanoparticles with narrow size distribution.

Transmission electron microscopy (TEM) was employed to determine the Ag nanoparticles' size distribution on CeO$_2$ surface. As shown in Fig. 2.17, the prepared Ag nanoparticles were mostly spherical in shape and the particle size positively correlated with the irradiation time. As the irradiation time increases, the average particle size of Ag nanoparticles increases; however, in the time between 120 and 160 s, the Ag precursors were almost consumed, which nearly stopped the growth of Ag nanoparticles. After 180 s, the Ag particles tended to aggregate, and the particle size grew. This concludes that the irradiation time between 120 and 160 s could be considered appropriate for the sufficient reduction of Ag$^+$ nanoparticles while preventing the particles from aggregating. It can be concluded that the shorter reaction times of microwave treatment generated controllable smaller sized Ag nanoparticles on the surface of CeO$_2$. The catalytic activity results showed that the Ag/CeO$_2$ exhibits good CO oxidation with overall conversion at 200°C.

Other important catalysts, such as Ni-CaO, Ni-Al$_2$O$_3$, BiVO$_4$, and so on, were also prepared using MWCM for various catalytic applications (Cesário *et al.*, 2015; Gao *et al.*, 2016; Abraham *et al.*, 2016).

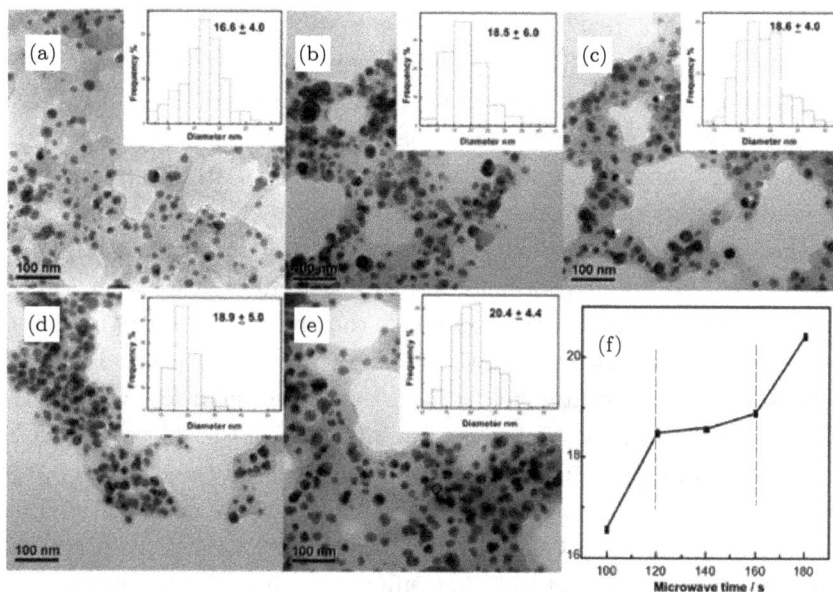

Figure 2.17. TEM images of Ag colloids with increasing irradiation time: (a) 100 s, (b) 120 s, (c) 140 s, (d) 160 s, (e) 180 s, and (f) particle mean size *vs.* irradiation time. (Reprinted with permission from Yang *et al.* (2015). Copyright © Elsevier.)

2.4. Conclusion

In this chapter, the principles behind microwave treatment and the synthesis of various nanostructured catalysts through the MWCM were summarized. The SCS has proved to be one of the most scalable and economical technologies for preparing nanostructured catalysts over the past two decades. It has enabled us to control different parameters such as pH, fuel type and amount, type of metal precursors, and external templates, which all affect the morphology, texture, and redox properties of the catalyst. Besides, compared to the conventional heating methods, microwave technology has become an acceptable technology for the preparation of desired nanostructured catalysts. The combination of these individual technologies often offer several advantages such as (1) fast reaction time and high production yield, (2) highly controllable and can be on and off within short reaction times, eliminating the need for warm-up and cool-down, (3) easy, rapid, energy-saving, and ready-to-obtain homogeneously dispersed spherical nanoparticles, (4) high potentiality and versatility, (5) capable of growing with the existing technologies, and (6) new materials optimization strategies are also possible. Further, the other aspects of the microwave-heating-SCS method include a high purity of nanostructured catalysts, the usual absence of unwanted products, and its sustainable character, all of which find a suitable collocation inside the green chemistry and green engineering fields too. In addition, the SCS requires relatively cheap reactants, simple equipment, and cost-effective facilities, which further strengthen the possible commercialization of the microwave-SCS method to industrial scale. It is noted that most of the catalysts prepared using this method exhibit high surface area and porosity while compared to the other CCMs. The catalytic activities of the prepared materials using this method offer enhanced catalytic performance.

However, the SCS method has few limitations when metal nitrates are used as precursors during the combustion reaction. Metal precursors can undergo partial thermal oxidation, which can generate potential hazardous gases like NO_x (NO_2, NO, and N_2O) that could form in large scale when the method was employed in pilot scale as well as industrial level. The emission of NO_x gases may be controlled if specific selective catalytic reduction (SCR) plant is installed in industries. Precise control over this technology by changing the fuel to oxidizer ratio may reduce these NO_x emissions. Another disadvantage is the formation of carbon if excessive fuel is used during the synthesis. Again, high temperatures are required to remove this carbon, which obviously decreases the surface area of the catalyst. This example clearly demonstrates the importance of proper fuel selection. Another limitation is the usage of the instrumental apparatus and understanding the effect of different parameters such as irradiation time and microwave power. Designing controllable and easily operated instruments could make the process more feasible in industries.

Finally, the microwave-assisted SCS method has proved to be an economic, rapid, and energy-saving route for nanostructured catalysts. The advantages of shorter setup and commissioning time and easy manufacturing during a shorter time and saving electrical energy could make this technology possible in industrial scale. However, controlling exhaust gases, stopping heat transfer to the surroundings, reproducibility and control over process quality are considered risks in large scale. Nowadays, safety has become one more important criterion in industries. Therefore, by taking into consideration all these factors, there is a possibility of using microwave-assisted synthesis approach for making industrial scale catalysts. Further investigations are necessary to explore this method for the synthesis of various new materials useful for numerous applications. For example, this methodology has been capable of growing with technology and development and it is expected that it will continue growing in the future, with new approaches and strategies for the materials preparation, particularly, a sustainable transformation of waste-derived and natural precursors into valuable products.

Acknowledgments

We wish to specially acknowledge all the researchers, whose works are described in this chapter for their valuable contributions. Benjaram M. Reddy thanks the Department of Atomic Energy (DAE), Mumbai, for the award of Dr Raja Ramanna Fellowship.

References

Abraham, S. D., David, S. T., Bennie, R. B., Joel, C., & Kumar, D. S. (2016). Eco-friendly and green synthesis of $BiVO_4$ nanoparticle using microwave irradiation as photocatalyst for the degradation of alizarin red s. *J. Molec. Struct., 1113*, 174–181.

Ajamein, H., Haghighi, M., Alaei, S., & Minaei, S. (2017). Ammonium nitrate-enhanced microwave solution combustion fabrication of $CuO/ZnO/Al_2O_3$ nanocatalyst for fuel cell grade hydrogen supply. *Microporous Mesoporous Mater., 245*, 82–93.

Anastas, P. T. (2007). Introduction: Green chemistry. *Chem. Rev., 107*, 2167–2168.

Asefi, N., Masoudpanah, S. M., & Hasheminiasari, M. (2018). Microwave assisted solution combustion synthesis of $BiFeO_3$ powders. *J. Sol-Gel Sci. Technol., 86*, 751–759.

Athayde, D. D., Souza, D. F., Silva, A. M. A., Vasconcelos, D., Nunes, E. H. M., Diniz da Costa, J. C., & Vasconcelos, W. L. (2016). Review of perovskite ceramic synthesis and membrane preparation methods. *Ceram. Int., 42,* 6555–6571.

Baig, R. B. N., & Varma, R. S. (2012). Alternative energy input: Mechanochemical, microwave and ultrasound-assisted organic synthesis. *Chem. Soc. Rev., 41,* 1559–1584.

Baláž, P., Godočíková, E., Kril'ová, L., Lobotka, P., & Gock, E. (2004). Preparation of nanocrystalline materials by high-energy milling. *Mater. Sci. Eng., A, 386,* 442–446.

Barros, B. S., Melo, D. M. A., Libs, S., & Kiennemann, A. (2010). CO2 reforming of methane over $La_2NiO_4/\alpha–Al_2O_3$ prepared by microwave assisted self-combustion method. *Appl. Catal. A, 378,* 69–75.

Buha, J., Djerdj, I., & Niederberger, M. (2007). Nonaqueous synthesis of nanocrystalline indium oxide and zinc oxide in the oxygen-free solvent acetonitrile. *Cryst. Growth Des., 7,* 113–116.

Cao, Y., Liu, B., Huang, R., Xia, Z., & Ge, S. (2011). Flash synthesis of flower-like ZnO nanostructures by microwave-induced combustion process. *Mater. Lett., 65,* 160–163.

Carreno, N. L. V., Deon, V. G., Silva, R. M., Santana, L. R., Pereira, R. M., Orlandi, M. O., ... Mesko, M. F. (2018). Feasible and clean solid-phase synthesis of LiNbO3 by microwave-induced combustion and its application as catalyst for low-temperature aniline oxidation. *ACS Sustainable Chem. Eng., 6,* 1680–1691.

Cesário, M. R., Barros, B. S., Courson, C., Melo, D. M. A., & Kiennemann, A. (2015). Catalytic performances of Ni–CaO–Mayenite in CO2 sorption enhanced steam methane reforming. *Fuel Process. Technol., 131,* 247–253.

Chang, W., Skandan, G., Danforth, S. C., Kear, B. H., & Hahn, H. (1994). Chemical vapor processing and applications for nanostructured ceramic powders and whiskers. *Nanostruct. Mater., 4,* 507–520.

Chen, C.-W., Riman, R. E., TenHuisen, K. S., & Brown, K. (2004). Mechanochemical–hydrothermal synthesis of hydroxyapatite from nonionic surfactant emulsion precursors. *J. Cryst. Growth, 270,* 615–623.

Cheng, P., Li, W., Liu, H., Gu, M., & Shangguah, W. (2004). Influence of zinc ferrite doping on the optical properties and phase transformation of titania powders prepared by sol–gel method. *Mater. Sci. Eng.: A, 386,* 43–47.

Choi, J.-H., Park, K.-W., Kwon, B.-K., & Sung, Y.-E. (2003). Methanol oxidation on Pt/Ru, Pt/Ni, and Pt/Ru/Ni anode electrocatalysts at different temperatures for DMFCs. *J. Electrochem. Soc., 150,* A973–A978.

Da Silveira, V. R., Melo, D. M. A., Barros, B. S., Ruiz, J. A. C., & Rojas, L. O. A. (2016). Nickel-based catalyst derived from $NiO–Ce_{0.75}Zr_{0.25}O_2$ nanocrystalline composite: Effect of the synthetic route on the partial oxidation of methane. *Ceram. Int., 42,* 16084–16089.

Deganello, F., & Tyagi, A. K. (2018). Solution combustion synthesis, energy and environment: Best parameters for better materials. *Prog. Cryst. Growth Charact. Mater., 64,* 23–61.

Devaiah, D., Reddy, L. H., Park, S. -E., & Reddy, B. M. (2018). Ceria–zirconia mixed oxides: Synthetic methods and applications. *Catal. Rev., 60*, 177–277.

Ede, S. R., Anantharaj, S., Subramanian, B., Rathishkumar, A., & Kundu, S. (2018). Microwave-assisted template-free synthesis of $Ni_3(BO_3)_2$ (NOB) hierarchical nanoflowers for electrocatalytic oxygen evolution. *Energy Fuels, 32*, 6224–6233.

El Khaled, D., Novas, N., Gazquez, J. A., & Manzano-Agugliaro, F. (2018). Microwave dielectric heating: Applications on metals processing. *Renewable Sustainable Energy Rev., 82*, 2880–2892.

Esmaeilnejad-Ahranjani, P., Khodadadi, A., Ziaei-Azad, H., & Mortazavi, Y. (2011). Effects of excess manganese in lanthanum manganite perovskite on lowering oxidation light-off temperature for automotive exhaust gas pollutants. *Chem. Eng. J., 169*, 282–289.

Fan, X., Li, L., Jing, F., Li, J., & Chu, W. (2018). Effects of preparation methods on $CoAlO_x/CeO_2$ catalysts for methane catalytic combustion. *Fuel, 225*, 588–595.

Fini, A., & Breccia, A. (1999). Chemistry by microwaves. *Pure and Appl. Chem., 71*, 573–579.

Gao, Y., Meng, F., Ji, K., Song, Y., & Li, Z. (2016). Slurry phase methanation of carbon monoxide over nanosized $Ni–Al_2O_3$ catalysts prepared by microwave-assisted solution combustion. *Appl. Catal. A, 510*, 74–83.

Gedye, R., Smith, F., Westaway, K., Ali, H., Baldisera, L., Laberge, L., & Rousell, J. (1986). The use of microwave ovens for rapid organic synthesis. *Tetrahedron Lett., 27*, 279–282.

Giguere, R. J., Bray, T. L., Duncan, S. M., & Majetich, G. (1986). Application of commercial microwave ovens to organic synthesis. *Tetrahedron Lett., 27*, 4945–4948.

Gillon, P., Courville, R., Steinchen, A., & Laliemant, M. (1987). Evaporation of polar liquids under low pressure and/or microwaveirradiation. *J. Micro. Power Electromagn. Energy, 22*, 155–166.

Gude, V. G., Patil, P., Martinez-Guerra, E., Deng, S., & Nirmalakhandan, N. (2013). Microwave energy potential for biodiesel production. *Sustainable Chem. Processes, 1*, 1–31.

Hashemzehi, M., Saghatoleslami, N., & Nayebzadeh, H. (2017). Microwave-assisted solution combustion synthesis of spinel-type mixed oxides for esterification reaction. *Chem. Eng. Commun., 204*, 415–423.

Hoffmann, R. C., Sanctis, S., Erdem, E., Weber, S., & Schneider, J. J. (2016). Zinc diketonates as single source precursors for zno nanoparticles: Microwave-assisted synthesis, electrophoretic deposition and field-effect transistor device properties. *J. Mater. Chem. C, 4*, 7345–7352.

Jacob, J., Chia, L. H. L., & Boey, F. Y. C. (1995). Thermal and non-thermal interaction of microwave radiation with materials. *J. Mater. Sci., 30*, 5321–5327.

Jesudoss, S. K., Vijaya, J. J., Kennedy, L. J., Rajan, P. I., Al-Lohedan, H. A., Ramalingam, R. J., ... Bououdina, M. (2016). Studies on the efficient dual performance of $Mn_{1-X}Ni_xFe_2O_4$ spinel nanoparticles in photodegradation and antibacterial activity. *J. Photochem. Photobiol., B, 165*, 121–132.

Kalantari, Z. B., Masoudpanah, S. M., & Hasheminiasari, M. (2016). Photocatalytic properties of ZnO powders synthesized by conventional and microwave-assisted solution combustion method. *J. Sol-Gel Sci. Technol., 3*, 711–718.

Kenji, K., Kanuma, Y., Oguro, T., & Harada, A. (1986). The reliability of magnetrons for microwave ovens. *J. Micro. Power Electromagn. Energy, 21*, 149–158.

Kharisov, B. I., Dias, H. V. R., & Kharissova, O. V. (2014). Mini-review: Ferrite nanoparticles in the catalysis. *Arabian J. Chem.* doi: 10.1016/j.arabjc.2014.10.049.

Kitchen, H. J., Vallance, S. R., Kennedy, J. L., Tapia-Ruiz, N., Carassiti, L., Harrison, A., ... Gregory, D. H. (2014). Modern microwave methods in solid-state inorganic materials chemistry: From fundamentals to manufacturing. *Chem. Rev., 114*, 1170–1206.

Kombaiah, K., Judith, J. V., John, L. K., & Kaviyarasu, K. (2019). Catalytic studies of $NiFe_2O_4$ nanoparticles prepared by conventional and microwave combustion method. *Mater. Chem. Phys., 221*, 11–28.

Kooti, M., & Naghdi Sedeh, A. (2013). Microwave-assisted combustion synthesis of ZnO nanoparticles. *J. Chem., 2013*, 562028–562031.

Köseoğlu, Y. (2014). A simple microwave-assisted combustion synthesis and structural, optical and magnetic characterization of ZnO nanoplatelets. *Ceram. Int., 40*, 4673–4679.

Li, Y., & Yang, W. (2008). Microwave synthesis of zeolite membranes: A review. *J. Memb. Sci., 316*, 3–17.

Liu, H., He, Z., Jiang, L.-P., & Zhu, J.-J. (2015). Microwave-assisted synthesis of wavelength-tunable photoluminescent carbon nanodots and their potential applications. *ACS Appl. Mater. Interfaces, 7*, 4913–4920.

Liu, Y.-C., & Fu, Y.-P. (2010). Magnetic and catalytic properties of copper ferrite nanopowders prepared by a microwave-induced combustion process. *Ceram. Int., 36*, 1597–1601.

Maghsoodi, S., Towfighi, J., Khodadadi, A., & Mortazavi, Y. (2013). The effects of excess manganese in nano-size lanthanum manganite perovskite on enhancement of trichloroethylene oxidation activity. *Chem. Eng. J., 215–216*, 827–837.

Mahmoodi, N. M., Masrouri, O., & Arabi, A. M. (2014). Synthesis of porous adsorbent using microwave assisted combustion method and dye removal. *J. Alloys Compounds, 602*, 210–220.

Mahmoud, M. H., Hassan, A. M., Said, A. E.-A. A., & Hamdeh, H. H. (2016). Structural; magnetic and catalytic properties of nanocrystalline $Cu_{0.5}Zn_{0.5}Fe_2O_4$ synthesized by microwave combustion and ball milling methods. *J. Molec. Struct., 1114*, 1–6.

Manikandan, A., Sridhar, R., Arul Antony, S., & Ramakrishna, S. (2014). A simple aloe vera plant-extracted microwave and conventional combustion synthesis: Morphological, optical, magnetic and catalytic properties of $CoFe_2O_4$ nanostructures. *J. Molec. Struct., 1076*, 188–200.

Manukyan, K. V. (2017). Solution combustion synthesis of catalysts. In I. P. Borovinskaya, A. A. Gromov, E. A. Levashov, Y. M. Maksimov, A. S. Mukasyan, & A. S. Rogachev (Eds.), *Concise Encyclopedia of Self-Propagating High-Temperature Synthesis* (347–348). Amsterdam: Elsevier.

McBride, J. R., Hass, K. C., Poindexter, B. D., & Weber, W. H. (1994). Raman and X-ray studies of $Ce_{1-X}Re_xO_{2-Y}$, where Re = La, Pr, Nd, Eu, Gd, and Tb. *J. Appl. Phy., 76*, 2435–2441.

Meher, S. K., & Rao, G. R. (2012). Polymer-assisted hydrothermal synthesis of highly reducible shuttle-shaped CeO_2: Microstructural effect on promoting Pt/C for methanol electrooxidation. *ACS Catal., 2*, 2795–2809.

Mingos, D. M. P., & Baghurst, D. R. (1991). Tilden Lecture: Applications of microwave dielectric heating effects to synthetic problems in chemistry. *Chem. Soc. Rev., 20*, 1–47.

Mirzaei, A., & Neri, G. (2016). Microwave-assisted synthesis of metal oxide nanostructures for gas sensing application: A review. *Sensors and Actuators B: Chemical, 237*, 749–775.

Mohammadzadeh, S., Olya, M. E., Arabi, A. M., Shariati, A., & Khosravi Nikou, M. R. (2015). Synthesis, characterization and application of ZnO–Ag as a nanophotocatalyst for organic compounds degradation, mechanism and economic study. *J. Environ. Sci., 35*, 194–207.

Nayebzadeh, H., Saghatoleslami, N., Haghighi, M., & Tabasizadeh, M. (2017). Influence of fuel type on microwave-enhanced fabrication of $KOH/Ca_{12}Al_{14}O_{33}$ nanocatalyst for biodiesel production via microwave heating. *J. Taiwan Inst. Chem. Eng., 75*, 148–155.

Ong, C. B., Ng, L. Y., & Mohammad, A. W. (2018). A review of ZnO nanoparticles as solar photocatalysts: Synthesis, mechanisms and applications. *Renewable and Sustainable Energy Rev., 81*, 536–551.

Padmapriya, G., Manikandan, A., Krishnasamy, V., Jaganathan, S. K., & Antony, S. A. (2016). Spinel $NixZn_{1-X}Fe_2O_4$ (0.0 ≤ x ≤ 1.0) nano-photocatalysts: Synthesis, characterization and photocatalytic degradation of methylene blue dye. *J. Molec. Struct., 1119*, 39–47.

Patil, K. C., Aruna, S. T., & Mimani, T. (2002). Combustion synthesis: An update. *Current Opinion in Solid State and Materials Science, 6*, 507–512.

Purohit, R. D., Saha, S., & Tyagi, A. K. (2001). Nanocrystalline thoria powders via glycine-nitrate combustion. *J. Nuc. Mater., 288*, 7–10.

Ragupathi, C., Vijaya, J. J., & Kennedy, L. J. (2014). Synthesis, characterization of nickel aluminate nanoparticles by microwave combustion method and their catalytic properties. *Mater. Sci. Eng., B, 184*, 18–25.

Ragupathi, C., Vijaya, J. J., Kennedy, L. J., & Bououdina, M. (2014). Nanostructured copper aluminate spinels: Synthesis, structural, optical, magnetic, and catalytic properties. *Mater. Sci. Semicond. Process., 24*, 146–156.

Rao, K. J., Vaidhyanathan, B., Ganguli, M., & Ramakrishnan, P. A. (1999). Synthesis of inorganic solids using microwaves. *Chem. Mater., 11*, 882–895.

Reddy, B. M., Reddy, G. K., Rao, K. N., Ganesh, I., & Ferreira, J. M. F. (2009). Characterization and photocatalytic activity of TiO_2–M_xO_y ($M_xO_y = SiO_2$, Al_2O_3, and ZrO_2) mixed oxides synthesized by microwave-induced solution combustion technique. *J. Mater. Sci., 44*, 4874–4882.

Reddy, G. K., Thrimurthulu, G., & Reddy, B. M. (2009). A rapid microwave-induced solution combustion synthesis of ceria-based mixed oxides for catalytic applications. *Catal. Surv. Asia, 13*, 237–255.

Reddy, L. H., Devaiah, D., & Reddy, B. M. (2014). Microwave assisted synthesis: A versatile tool for process intensification. In K. V. Raghavan & B. M. Reddy (Eds.), *Industrial catalysis and separations* (pp 375–405, chap. 10). USA: Apple Academic Press, Inc.

Reddy, L. H., Reddy, G. K., Devaiah, D., & Reddy, B. M. (2012). A rapid microwave-assisted solution combustion synthesis of CuO promoted CeO_2–M_xO_y (M = Zr, La, Pr and Sm) catalysts for CO oxidation. *Appl. Catal. A, 445–446*, 297–305.

Rezaee, L., & Haghighi, M. (2016). Citrate complexation microwave-assisted synthesis of $Ce_{0.8}Zr_{0.2}O_2$ nanocatalyst over Al_2O_3 used in CO oxidation for hydrogen purification: Influence of composite loading and synthesis method. *RSC Adv., 6*, 34055–34065.

Rodnyi, P. A., & Khodyuk, I. V. (2011). Optical and luminescence properties of zinc oxide (review). *Opt. Spectrosc., 111*, 776–785.

Rogachev, A. S., & Baras, F. (2007). Models of shs: An overview. *Int. J. Self-Propagating High-Temperature Synthesis, 16*, 141–153.

Rosa, R., Veronesi, P., & Leonelli, C. (2013). A review on combustion synthesis intensification by means of microwave energy. *Chem. Eng. Process. Process Intensification., 71*, 2–18.

Sahu, R. K., Ganguly, K., Mishra, T., Mishra, M., Ningthoujam, R. S., Roy, S. K., & Pathak, L. C. (2012). Stabilization of intrinsic defects at high temperatures in ZnO nanoparticles by Ag modification. *J. Colloid Interface Science, 366*, 8–15.

Samantaray, S., & Mishra, B. G. (2011). Combustion synthesis, characterization and catalytic application of MoO_3–ZrO_2 nanocomposite oxide towards one pot synthesis of octahydroquinazolinones. *J. Molec. Catal. A: Chem., 339*, 92–98.

Selvam, N. C. S., Kumar, R. T., Kennedy, L. J., & Vijaya, J. J. (2011). Comparative study of microwave and conventional methods for the preparation and optical properties of novel MgO-micro and nano-structures. *J. Alloys Compd., 509*, 9809–9815.

Shahmirzaee, M., Shafiee Afarani, M., Arabi, A. M., & Iran Nejhad, A. (2017). In situ crystallization of $ZnAl_2O_4$/ZnO nanocomposite on alumina granule for photocatalytic purification of wastewater. *Res. Chem. Intermed., 43*, 321–340.

Sherly, E. D., Vijaya, J. J., & Kennedy, L. J. (2015). Effect of CeO_2coupling on the structural, optical and photocatalytic properties of ZnO nanoparticle. *J. Molec. Struct., 1099*, 114–125.

Specchia, S., Ercolino, G., Karimi, S., Italiano, C., & Vita, A. (2017). Solution combustion synthesis for preparation of structured catalysts: A mini-review on process intensification for energy applications and pollution control. *Int. J. Self-Propagating High-Temperature Synthesis, 26*, 166–186.

Sundararajan, M., Sailaja, V., John Kennedy, L., & Judith Vijaya, J. (2017). Photocatalytic degradation of rhodamine b under visible light using nanostructured zinc doped cobalt ferrite: Kinetics and mechanism. *Ceram. Int., 43*, 540–548.

Tahmasebi, K., & Paydar, M. H. (2011). Microwave assisted solution combustion synthesis of alumina–zirconia, ZTA, nanocomposite powder. *J. Alloys Compd., 509*, 1192–1196.

van Eldik, R., & Hubbard, C. D. (1996). *Chemistry under Extreme and Non-Classical Conditions.* John Wiley & Sons.

Varma, A., Mukasyan, A. S., Rogachev, A. S., & Manukyan, K. V. (2016). Solution combustion synthesis of nanoscale materials, *Chem. Rev., 116*, 14493–14586.

Vie, D., Martínez, E., Sapiña, F., Folgado, J.-V., Beltrán, A., Valenzuela, R. X., & Cortés-Corberán, V. (2004). Freeze-dried precursor-based synthesis of nanostructured cobalt–nickel molybdates $Co_{1-x}Ni_xMoO_4$. *Chem. Mater., 16*, 1697–1703.

Wan, J., Huang, L., Wu, J., Xiong, L., Hu, Z., Yu, H., ... Zhou, J. (2018). Microwave combustion for rapidly synthesizing pore-size-controllable porous graphene. *Adv. Funct. Mater., 28*, 1800382–1800388.

Wang, X., Liu, D., Song, S., & Zhang, H., (2013). Pt@CeO2 multicores shell self-assembled nanospheres: Clean synthesis, structure optimization, and catalytic applications. *J. Am. Chem. Soc., 135*, 15864–15872.

Yang, F., Huang, J., Odoom-Wubah, T., Hong, Y., Du, M., Sun, D., Jia, L., & Li, Q. (2015). Efficient Ag/CeO2 catalysts for CO oxidation prepared with microwave-assisted biosynthesis. *Chem. Eng. J., 269*, 105–112.

Zawadzki, M., Walerczyk, W., López-Suárez, F. E., Illán-Gómez, M. J., & Bueno-López, A. (2011). CoAl2O4 spinel catalyst for soot combustion with NO_x/O_2. *Catal. Commun., 12*, 1238–1241.

© 2021 World Scientific Publishing Company
https://doi.org/10.1142/9781786348708_0003

Chapter 3

Solution Combustion Synthesis Related to Photocatalytic Reactions

Sounak Roy* and Swapna Challagulla[†]
*Department of Chemistry, BITS Pilani, Hyderabad Campus,
Jawahar Nagar, Shameerpet Mandal, Hyderabad 500078, India*
*sounak.roy@hyderabad.bits-pilani.ac.in
[†]challagulla.swapna@gmail.com

3.1. Introduction

The solution combustion synthesis (SCS) is the most promising, facile, energy-efficient, and economically viable method for the synthesis of a variety of nano-structured solid catalysts. The SCS is generally initiated at low temperatures and is a single-step quick synthetic process, which makes it an intense method for the synthesis of low-cost materials compared with that of other conventional methods. The advantages of the SCS are (1) requirement of low-temperature equipment, (2) high purity of formed products, (3) formation of stable phase, (4) even distribution of dopants on the materials, and (5) formation of controlled shape, homogeneous, and stoichiometric materials.

In the SCS, water-soluble metal nitrate precursors with preferred fuel such as glycine, urea, oxalic acid, carbohydrates, or hydrazides such as tetra formal trisazine (TFTA), oxalyldihydrazide (ODH), carbo hydrazide (CH), and malonic acid dihydrazide (MDH) are used for the preparation of catalytic materials. The factors that influence the exothermic SCS process are fuel-to-oxidizer ratio, amount of water in the precursor solution, and combustion temperature. The choice of fuels greatly influences the characteristics of the final product because different fuels have different reducing power, ignition temperature, and amount of released gases. The fuels with lower decomposition temperature are important to generate local heat for the complete combustion of precursor mixture and also help to form the nano-sized product with high crystallinity, high porosity, and uniform distribution. The ideal features of a fuel are (1) high solubility in water, (2) low combustion temperature, (3) un-explosive nature, (4) releasing harmless gases, and (5) having no residual after ignition. The fuel primarily has two important roles: one behaving as chelating agent and the other being template for microstructures. The fuel interacts with metal precursor to form a chelate complex with metal ions, which is necessary for effective combustion. This chelated complex of the metal is introduced for uncontrolled heating at or above the decomposition temperature of the fuel. During combustion, water is evaporated initially forming a sol, and eventually forms a gel network, which helps to maintain a homogeneous position for metal cations during combustion, and makes a proper connection between fuel and oxidant species (Figure 3.1).

The nature of the combustion is also influenced by the oxidizer-to-fuel ratio, which in turn influences the morphology, phase composition, surface properties, metal cations, oxidation states, oxygen vacancies, and other material properties (Li *et al.*, 2015; Wen and Wu, 2014) of the final solid product. The equivalence ratio of an oxidizer to fuel is calculated using the following equation:

$$\varphi_e = \frac{\Sigma \text{ Coefficient of oxdizing element in the formula} \times \text{Valency}}{(-1) \, \Sigma \text{ Coefficient of reducing element in the formula} \times \text{Valency}},$$

where φ_e = elemental stoichiometric equivalence coefficient. If $\varphi_e = 1$, then the mixture is in stoichiometric ratio. If $\varphi_e < 1$, then the mixture is said to be fuel-rich, and if $\varphi_e > 1$, it is called fuel-lean composition. During spontaneous combustion, the oxidizer and the fuel are ignited rapidly and decompose to gaseous products as foam and leads to huge swelling. The decomposed gaseous products are primarily the oxides of nitrogen (NO_2, NO), ammonia, CO_2, CO, and

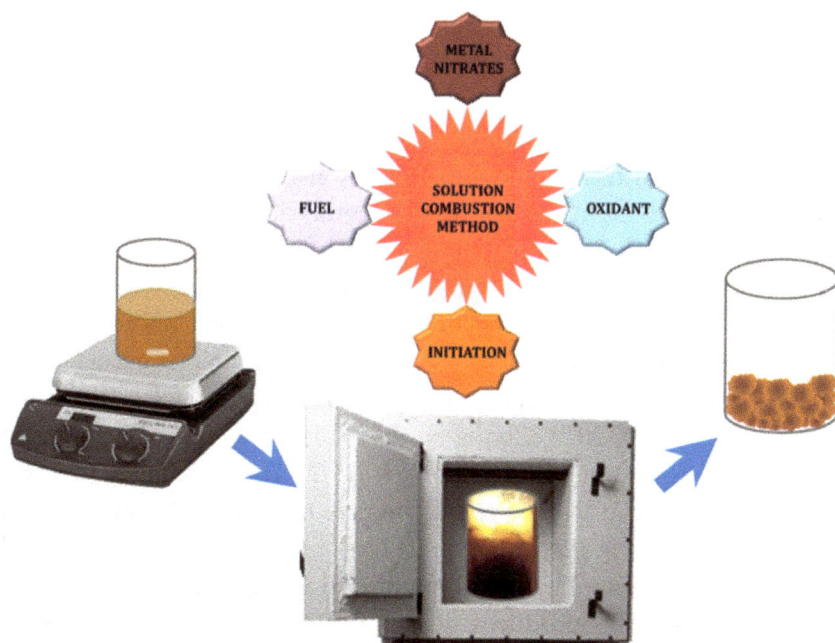

Figure 3.1. The four major "ingredients" of the SCS and step-by-step procedure.

HNCO, which are hypergolic. This hypergolic mixture of gases breaks the foam with flame and raises the in situ temperature above 1000°C. Subsequently, the whole foam burns with flame. The hypergolic gases are responsible for the complete combustion of the mixture to form nanostructured porous catalysts (Civera *et al.*, 2003; Tasca *et al.*, 2011). However, if heating follows a controlled heating rate to reach combustion or ignition temperature, then reaction mixture starts melting followed by dehydration and decomposition finally to form a solid product. This method is suitable to make supported catalysts (Mukasyan *et al.*, 2007; Rogachev and Muskasyan, 2014).

3.2. Photocatalysis

The SCS method in a single step can quickly process a wide range of semiconducting catalytic materials, which are the most promising candidates for dye degradation, water splitting, hydrogen production, solar cells, removal of organic pollutants, and so on. In photocatalysis, when semiconducting material absorbs light, it ejects electron from valence band to conduction band and leaves holes in the valance band. These electrons and the holes are the two species responsible for the photocatalytic activity of semiconducting materials; however, the recombination of electron in conduction band and hole in valance band may suppress the catalytic activity. To avoid the recombination, several researchers modified the materials by forming heterojunction between two materials or doping with any metals and nonmetals. The popular choices for the effective photocatalysts are semiconducting materials such as TiO_2 (Linsebigler *et al.*, 1995; Nakata *et al.*, 2012; Nakata and Fujidhima, 2012), ZnO (Ong *et al.*, 2018; Zhang *et al.*, 2014; Zhang *et al.*, 2015), WO_3 (Kim *et al.*, 2010; Kumar and Rao, 2017; Nagarjuna *et al.*, 2017; Xi *et al.*, 2012), and so on. By using the SCS method, the nanostructured photocatalytic materials, including metal nanoparticles, metal oxides, mixed metal oxides, perovskites, alloys, and supported catalysts, are extensively prepared. The SCS produces doped and supported catalysts with high surface area and smaller crystallite size of uniformly distributed particles. The shape, structure, size, specific surface area, and band gap of the materials influence the photocatalytic activity. The SCS method can greatly influence one or other properties of the synthesized materials. For example, during combustion, C or N or S is generally doped in the synthesized materials from the fuels. These anion doping can help reduce the band gap to make the material's visible light active. In Section 3.3, we will discuss the influence of the SCS over the activity of the popular photocatalysts.

3.3. Photocatalytic Materials by the SCS

3.3.1. *Titanium Dioxide*

Titanium dioxide (TiO_2) is one of the most promising semiconducting materials because of its chemically stable, low-cost, and nontoxic nature. Among the three polymorphic phases exhibited by TiO_2, anatase shows the highest catalytic activity (Hanaor and Sorrell, 2011; Luttrell *et al.*, 2014; Selcuk and Selloni, 2016). The intrinsic band gap of anatase TiO_2 is $\sim$3.2 eV, which limits its application in the UV range of light. Doping of TiO_2 with metals or nonmetals to decrease the band gap enhances the visible light absorption. The SCS can effectively alter the band gap, doping of metal or nonmetals on TiO_2 to enhance its efficiency toward various photocatalytic reactions.

The effect of type of fuel and the ratio between the fuel to the oxidizer on the structural and morphological properties of SCS-synthesized materials were thoroughly explained (Challagulla and Roy, 2017). It was observed that alteration from fuel-lean to stoichiometric to fuel-rich condition had changed the ratio between anatase and rutile phase of TiO_2. Three different fuels such as urea, glycine, and ODH were used for this study. Titanyl nitrate $(TiO(NO_3)_2)$ was taken as Ti source, and it was prepared hydrolyzing titanium (IV) isopropoxide with water followed by nitration with concentrated HNO_3. The reducing valences with three different fuels with three different conditions — fuel-lean, stoichiometric, and fuel-rich — were calculated according to propellant chemistry. The reducing and oxidizing valences of elements are considered as $C = +4$, $H = +1$, $O = -2$, $N = 0$, metal $= +2, +3$, and so on. For divalent metal nitrates, the oxidizing valency is -10, and for trivalent metal nitrates, the valency is -15. The oxidizing valency of metal nitrates was balanced by reducing valences of fuels. The reducing valences of urea, glycine, and ODH are $+6$, $+9$, and $+10$, respectively. The balanced stoichiometric equations $TiO(NO_3)_2$ with three different fuels are as follows:

1. With glycine:

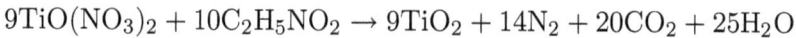
$$9TiO(NO_3)_2 + 10C_2H_5NO_2 \rightarrow 9TiO_2 + 14N_2 + 20CO_2 + 25H_2O$$

2. With urea:

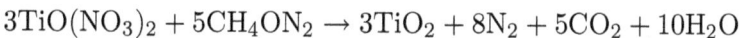
$$3TiO(NO_3)_2 + 5CH_4ON_2 \rightarrow 3TiO_2 + 8N_2 + 5CO_2 + 10H_2O$$

3. With ODH:

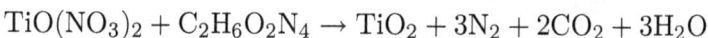
$$TiO(NO_3)_2 + C_2H_6O_2N_4 \rightarrow TiO_2 + 3N_2 + 2CO_2 + 3H_2O$$

In X-ray diffraction patterns, the TiO_2 synthesized with glycine exhibited pure anatase phase in all the three conditions — fuel-lean, stoichiometric, and fuel-rich — and formed mixed phase (anatase and rutile) when urea was used as fuel. There was no significant difference observed in the peak ratios between anatase and rutile from urea-lean to stoichiometric to urea-rich condition. On the contrary, TiO_2 made from ODH also formed mixed phase; however, the peak intensities of anatase to rutile had increased from ODH-lean to stoichiometric to ODH-rich condition. The combustion temperature significantly depended on the fuel and fuel-to-oxidizer ratio, which might have influenced the phase formation of TiO_2. Rutile TiO_2 is thermodynamically more stable than anatase, thus at higher combustion temperature rutile phase was observed. The anatase-to-rutile ratio of TiO_2 made with urea was found to be 89:11 in all the three conditions. This ratio of anatase to rutile was found to be 92:8, 88:12, and 84:16 for ODH-lean, stoichiometric, and ODH-rich conditions, respectively (Fig. 3.2A). Interestingly, the difference in the fuel did not influence the crystallite size. It was observed from field emission-scanning electron microscopy that TiO_2 made with glycine had spherical particles in nano range, whereas the TiO_2 made with urea and ODH had irregular rock-like structures. This clearly showed the effect of fuel on the surface morphology of combusted product (Fig. 3.2B). The mixed-phase TiO_2 showed high photocatalytic degradation of methylene blue than pure anatase TiO_2 and could be because of the formation of heterojunction between anatase and rutile phase. This minimized the electrons from recombining with holes, which was eventually effective for dye degradation. However, the existence of small particle size in TiO_2 made with glycine showed highest photocatalytic H_2 production when compared with TiO_2 made with urea and ODH. The effect of different fuels was also addressed by Nagaveni and coworkers, who synthesized TiO_2 with three different fuels: glycine, hexamethylenetetramine, (HMT), and ODH (Nagaveni *et al.*, 2004). The stoichiometric ratio of $TiO(NO_3)_2$ and fuel was combusted at $650°C$, which produced crystalline TiO_2. Interestingly, the color of TiO_2 made with glycine was found to be yellow, whereas white-color

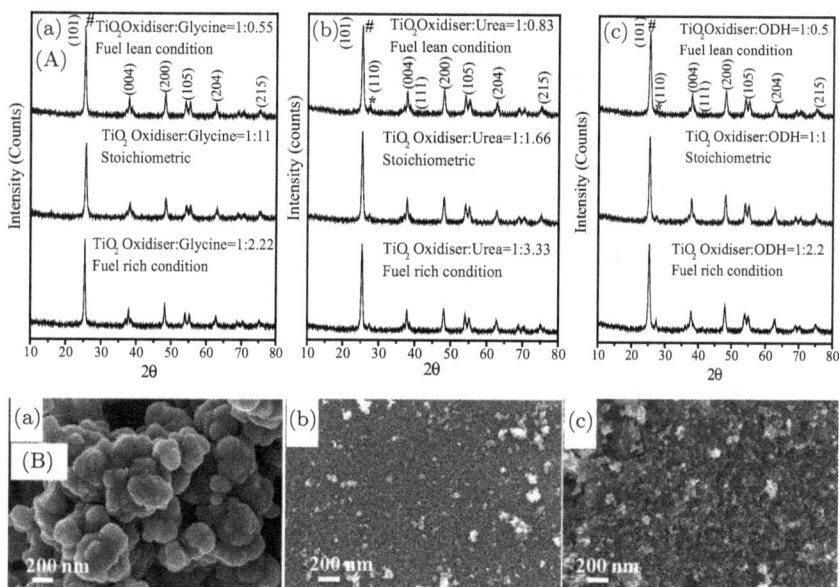

Figure 3.2. (A) X-ray powder diffraction (XRD) patterns and (B) field emission-scanning electron microscopy (FE-SEM) images of (a) TiO₂ with glycine, (b) TiO₂ with urea, and (c) TiO₂ with ODH. (# and * indicate anatase and rutile phases, respectively). Reproduced from Challagulla and Roy (2017) with copyright permission (Cambridge University Press).

TiO$_2$ was formed when HMT and ODH were used as fuels. It was proposed that the carbon and the nitrogen from the fuel glycine were doped on TiO$_2$ during combustion with glycine, which created oxide ion vacancy, and lowered the band gap of the formed yellow TiO$_2$. Thus, the catalyst showed absorption in visible region and also generated more number of surface hydroxyl groups, the key species for the catalytic degradation of methylene blue and phenol solar light irradiation. The surface area of synthesized catalysts was also found to be higher compared to the commercial Degussa P-25. Consequently, the combustion-synthesized TiO$_2$ showed higher rate of photodegradation of methylene blue and phenol when compared with Degussa P-25 (Fig. 3.3). The TiO$_2$ synthesized with the help of fuel glycine was extensively studied for other dyes and phenols, which also showed a very high activity of the material (Nagaveni *et al.*, 2004; Nagaveni *et al.*, 2004; Sivalingam and Madras, 2004;

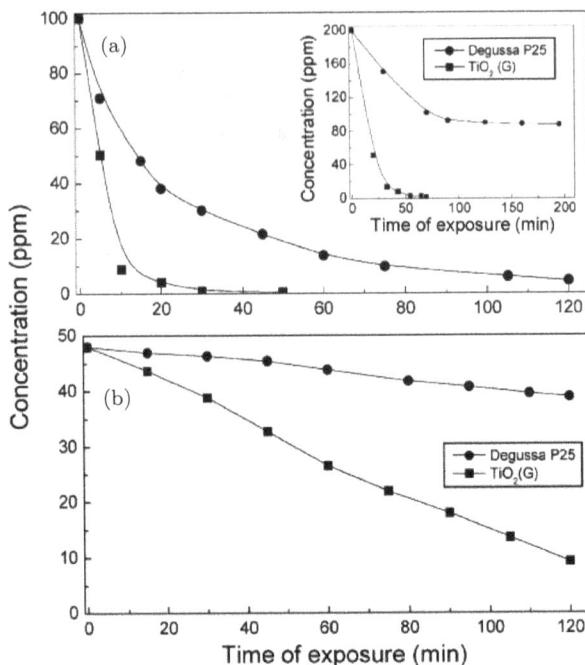

Figure 3.3. Degradation profiles of (a) methylene blue and (b) phenol over combustion-synthesized TiO₂ (TiO₂(G)) and Degussa P-25. Reproduced from Nagaveni *et al.* (2004) with copyright permission.

Sivalingam *et al.*, 2004). The TiO₂ synthesized using the SCS showed significant peak for both carbon and nitrogen in X-ray photoelectron spectroscopy (Mani *et al.*, 2012). A comparative study showed that among the three different synthetic methods of sol-gel, the SCS, and microwave-assisted hydrothermal method, the SCS produced anatase phase of TiO₂, which could efficiently produce hydrogen from photooxidation of methanol (Challagulla *et al.*, 2017) and could reduce nitrobenzene to aniline because of the synergistic effect between conduction band electrons of TiO₂ and onset reduction potential of nitrobenzene (Challagulla *et al.*, 2017). Sivaranjani and Gopinath (2011) synthesized wormhole mesoporous N-doped TiO₂ using the SCS method. The synthesized material exhibited visible light absorption, high surface area, small crystallite size, wormhole mesoporous network, and anatase phase. These properties enhanced the catalytic activity toward rhodamine B photodegradation and photocatalytic oxidation of aqueous p-anisyl alcohol to p-anisaldehyde in visible light. Pany and coauthors also synthesized wormhole mesoporous N-doped TiO_2, but modified with carbon and sulfur by sol-gel auto-combustion method and used for the degradation of phenol under visible light exposure (Pany *et al.*, 2013). Here, N incorporation improved the absorption in visible region, and S doping enhanced the electron hole separation and the presence of C on surface acts as sensitizer for transporting electrons through heterojunction. All these features enhanced the photocatalytic activity of combustion-synthesized N-doped TiO_2 modified with carbon and sulfur.

The noble, transition, and rare-earth metal-doped TiO_2 synthesized using the SCS method was also studied for a wide range of photocatalytic applications. The noble and transition metals were found to be existing either in the host TiO_2 lattice as a substitution of Ti or in very highly dispersed state on the surface. The combustion-synthesized materials were compared with its analogues made by impregnation method and it was observed that the combustion-synthesized materials outperformed the impregnated materials. Pd-doped TiO_2 was synthesized using solution combustion for the photocatalytic NO reduction and CO oxidation (Roy *et al.*, 2007). The substitution of Pd^{+2} ion on TiO_2 helped creating oxide ion vacancies and redox adsorption sites. The high conversion of NO was observed with solution combustion-synthesized $Ti_{0.99}Pd_{0.01}O_{1.99}$ in the presence of CO when compared with 1% Pd/TiO_2 synthesized using impregnation method. The authors also studied the kinetics of the photocatalytic reduction of NO by CO using solution combustion-synthesized $Ti_{0.99}Pd_{0.01}O_{1.99}$. Oxide ion vacancies created using the SCS enhanced the photodissociation of NO with high rate (Roy *et al.*, 2007). Sn-doped TiO_2 was prepared using the SCS by varying the Sn loading in the range of 0%–20% (Bhange *et al.*, 2016). The catalysts were assessed for the photodegradation of methylene blue. The recombination of electrons and holes was suppressed by doping of Sn on TiO_2. The maximum loading was found to be 10% of Sn on the TiO_2 lattice. Thind *et al.* (2012) synthesized mesoporous N, W co-doped TiO_2 by the SCS method using urea as fuel (which is also an

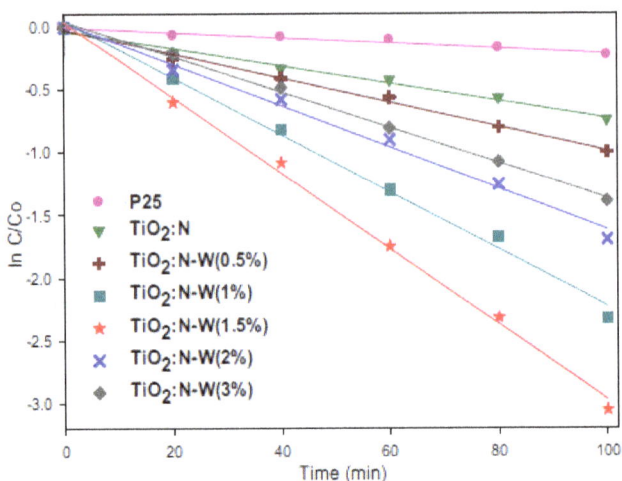

Figure 3.4. Kinetics curves for the photodegradation of rhodamine B over N- and W-doped combustion TiO$_2$ and P-25. Reproduced from Thind *et al.* (2004) with copyright permission.

N source), and sodium tungstate as W source. The characterization of the synthesized catalysts revealed the materials were mesoporous with high surface area. The doping decreased the band gap of synthesized materials and exhibited high absorption in visible region. The synthesized materials were investigated for the photocatalytic degradation of rhodamine B under solar light. The synergistic effect between visible light absorption and high surface area of combustion-derived N- and W-co-doped TiO$_2$ enhanced the visible light activity. The kinetic data for the degradation of rhodamine B over combustion synthesized TiO$_2$ and P-25. Figure 3.4 apparently shows the superiority of the combustion-synthesized material. Xiao *et al.* (2007) synthesized nanocrystalline Sm^{3+}-doped TiO$_2$ by sol-gel auto-combustion method and showed high photocatalytic activity in visible light. The doping of samarium on TiO$_2$ enhanced the absorption in visible region, which was confirmed by UV–Vis diffuse reflectance spectra. The synthesized materials employed for the degradation of methylene blue in solar light and 0.5 mol% Sm^{3+} doped on TiO$_2$ showed high catalytic efficiency. High activity of 0.5 mol% Sm^{3+} doped on TiO$_2$ catalyst was found because of the presence of higher oxygen vacancies and lower recombination rate of electrons and holes. A contradictory observation was reported when the SCS-synthesized different metal ions such as W, V, Ce, Zr, Fe, and Cu were doped on TiO_2 and screened for the degradation of 4-nitrophenol (Nagaveni *et al.*, 2004). Interestingly, the bare TiO_2 showed highest rate of degradation than metal-doped TiO_2. Similar observation was also corroborated by Vinu and Madras (2008), who synthesized Pd substituted and Pd impregnated on the nano TiO_2 by the SCS method, and screened their photocatalytic applicability. Unsubstituted TiO_2 showed higher catalytic efficiency than Pd-impregnated TiO_2 because of the decrease in the surface area and photoluminescence intensity on substituted TiO_2. However, Pd-substituted TiO_2 showed higher photocatalytic efficiency compared to the unsubstituted TiO_2. The difference of catalytic activity between the noble metal being substituted in TiO_2 matrix and dispersed over TiO_2 surface is also being highlighted in other references (Vinu and Madras, 2009). Ag/TiO_2 was synthesized using the SCS and used for the photocatalytic inactivation of *Escherichia coli* in visible light. The combustion-synthesized catalysts showed highest photocatalytic inactivation than commercial TiO_2 (Sontakke *et al.*, 2012).

The above discussion supports the superiority of the solution combustion-synthesized TiO_2 compared to the commercial one. However, a compendium of the catalytic reaction conditions and results is very much required. Table 3.1 is a compilation of the reaction condition used for the combustion-synthesized TiO_2.

The effect of adiabatic temperature and enthalpy on the phase transition in the combustion-synthesized TiO_2 as a function of fuel content was studied (Ma *et al.*, 2015). TiO_2 was synthesized using $TiO(NO_3)_2$ as oxidizer and urea as fuel, and varying the fuel-to-nitrate molar ratio from 0.25, 0.5, 0.6, 0.75, 1.0, and 1.5. The percentage of rutile phase increased with the increase in the fuel-to-nitrate ratio (0.25–0.75). TiO_2 exhibited 66.2% of anatase and 33.8% of rutile form when the molar ratio between urea and nitrate was 0.5, and showed high photocatalytic activity toward methyl orange degradation (80.1 % in 5 h). Heterojunction formed between anatase and rutile phase led to the fast charge separation between electrons

Table 3.1. Photocatalytic applications of the SCS of TiO_2, the reaction conditions, and inference.

Catalyst	Synthesis parameters	Reaction conditions	Inference	Reference
TiO_2	$TiO(NO_3)_2$ as Ti precursor, urea, glycine, and ODH as fuels, combustion at $450°C$	125-W medium pressure mercury vapor lamp, 100 mL of 10 ppm methylene blue solution, and 25 mg catalyst	Rate: 7.3×10^{-7} mol/(L-min) for TiO_2(urea), 2.7×10^{-7} mol/(L-min) for P-25	Challagulla and Roy (2017)
TiO_2	$TiO(NO_3)_2$ as Ti precursor, glycine, hexamethylenetetramine, and ODH as fuels, combustion at $350°C$	Sunlight with solar intensity $0.753 \, kW \, m^{-2}$, 100 ppm methylene blue solution, $1 \, kg \, m^{-3}$ catalyst	Initial rate: $0.153 \, \mu$ mol $L^{-1} s^{-1}$ for combustion TiO_2, $0.018 \, \mu$ mol $L^{-1} s^{-1}$ for Degussa P-25	Nagaveni *et al.* (2004)
TiO_2	$TiO(NO_3)_2$ as Ti precursor, glycine as fuel, combustion at $350°C$	125-W medium pressure mercury vapor lamp, 100 ppm phenol, $1 \, kg \, m^{-3}$ catalyst	Combustion TiO_2 showed higher rate of photodegradation than Degussa P-25	Sivalingam *et al.* (2014)
$TiO_{2-x}N_x$	Titanium nitrate as Ti source, urea as fuel, combustion at $400°C$	Visible light ($\lambda >$ $420 \, nm$), 100 ppm of rhodamine B, 100 mg of catalyst	$TiO_{2-x}N_x$ degraded 90% in 60 min. P-25 degraded $\sim$10% in 60 min	Sivaranjani and Gopinath (2011)

Table 3.1. (*Continued*)

Catalyst	Synthesis parameters	Reaction conditions	Inference	Reference
Sn-doped TiO$_2$	Titanium tetraiso-propoxide as Ti source, SnCl$_4$ as Sn source, urea as a fuel, combustion at 450°C	400-W mercury vapor lamp for UV light, 230 mL of 10 ppm methylene blue, and 23 mg of catalyst	Photocatalytic activity increases with increase in Sn content	Bhange *et al.* (2016)
Sm^{3+}-doped TiO$_2$	Titanium isopropoxide as Ti source, samarium nitrate as Sm source, citric acid as fuel, combustion at 250°C	150-W halogen tungsten lamp, 200 mL of 100 ppm methylene blue, 200 mg of catalyst	0.5% Sm^{3+}-doped TiO$_2$ showed 67.68% methylene blue at 120 min	Xiao *et al.* (2007)
N- and W-n co-doped TiO$_2$	Titanium (IV) isopropoxide as Ti source, urea as N source/fuel, and sodium tungstate as W source, combustion at 300°C	300-W xenon arc lamp, 20 mL of 25 μM rhodamine B, and 20 mg of catalyst	Rate constant: 3.01×10^{-2} min^{-1} for TiO$_2$:N-W (1.5%) and 2.12×10^{-3} min^{-1} for P-25	Thind *et al.* (2012)

and holes, and suppressed the recombination of electrons and holes, which were responsible for the high photocatalytic activity of the catalyst. This mechanism took place when the optimum amount of rutile phase formed a proper alignment with the anatase phase. The thermodynamic calculations with different urea to nitrate are shown in Fig. 3.5. Increase in the fuel ratio increased the adiabatic combustion temperature and enthalpy, thus favoring the increase

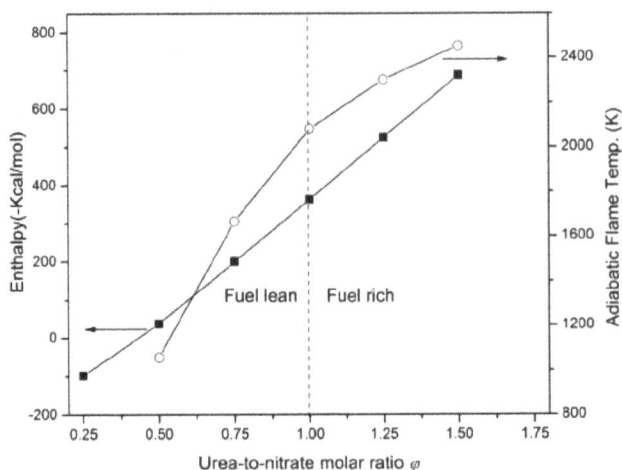

Figure 3.5. Variation of adiabatic flame temperature and enthalpy as a function of the amount of fuel used. Reproduced from Ma *et al.* (2015) with copyright permission (Elsevier).

of rutile phase. However, further increase in the flame adiabatic temperature did not completely remove the anatase because of the short time duration of combustion. The calcination of combustion-synthesized catalysts showed pure rutile phase.

3.3.2. *Zinc Oxide*

Other than TiO_2, zinc oxide (ZnO) is the second most promising semiconductor photocatalytic material with high thermal stability and wide band gap of $\sim$3.37 eV. The hexagonal wurtzite-structured ZnO and ZnO-based nanostructured materials had been extensively synthesized using the SCS method and were studied for a wide range of photocatalytic applications (Adhikari *et al.*, 2016; Assi *et al.*, 2015). Vasei and coworkers synthesized ZnO using the mixture of cetyltrimethylammonium bromide (CTAB) with glycine and citric acid as fuel in the SCS method (Vasei *et al.*, 2019). Though the fuel-to-oxidizer ratio was varied from 0.5, 0.75, 1.0, 1.5, the bulk structure of ZnO was not altered with respect to fuel or the fuel-to-oxidizer ratio. However, the electronic structure was influenced by the nature of the fuel used. Interestingly, the combination of CTAB with glycine formed a lower band gap, nano-crystallite sized ZnO, whereas CTAB with citric acid fuel formed high surface area and porous ZnO. It should be noted that the adiabatic flame temperature of CTAB and glycine (2100 K) is lower compared to the mixture of CTAB with citric acid fuel (2200 K). As-combusted ZnO materials were explored for the photocatalytic degradation of methylene blue, and ZnO formed by the combination of CTAB with citric acid because of its higher surface area showed higher activity. Recently, the same group also synthesized ZnO by varying hydrocarbon tail length of surfactants, which were used as fuel (Vasei *et al.*, 2019). Three different surfactants such as CTAB, dodecyltrimethylammonium bromide (DTAB), and octyltrimethylammonium bromide (OTAB) were used as fuels in the combustion process. Interestingly, the ZnO synthesized using longer hydrocarbon tail surfactant (CTAB) showed higher rate of photodegradation of methylene blue compared to others. ZnO was also synthesized by the SCS method using CTAB, where the effect of fuel-to-oxidizer ratio on the combustion behavior was studied (Vasei *et al.*, 2018). The ZnO with fuel-to-oxidizer ratio of 1.5 exhibited high rate of photodegradation of methylene blue because of the existence of good crystallinity and smaller particle size. Vasei's group also used polyvinylpyrrolidone and citric acid as fuels for the synthesis of honeycomb-like-structured ZnO, and used for the photocatalytic degradation of methylene blue (Vasei *et al.*, 2019). The influence of a wide range of fuels such as citric acid, dextrose, glycine, ODH, oxalic acid, and urea during the SCS of ZnO was studied (Potti and Srivastava, 2012), and FE-SEM images showed different surface morphology depending on the fuel (Fig. 3.6). ZnO with citric acid formed agglomerated particles because of the viscous nature of citric acid, whereas during combustion with dextrose, a large volume of gases escaped during combustion and formed highly porous and high surface area irregular-shaped ZnO. Spherical ZnO particles were formed when glycine and ODH were used as fuels. Cylindrical shape of ZnO particles was produced with oxalic acid, and urea formed flower-like structure with a very low porous structure. ZnO with oxalic acid as fuel showed high degradation and decolorization of orange dye among all other synthesized catalysts.

Figure 3.6. Scanning electron micrographs of ZnO synthesized using fuels: (a) citric acid, (b) dextrose, (c) glycine, (d) ODH, (e) oxalic acid, and (f) urea. Reproduced from Potti and Srivastava (2012) with copyright permission (American Chemical Society).

The influence of new category of fuels was studied when ZnO was synthesized from the combustion of a biofuel from tapioca starch pearls that are derived from tubers of Manihotesculenta (Ramasami *et al.*, 2015). The synthesized spherical particles of ZnO were formed with 2.7 eV band gap and with high porosity. The catalysts showed high methylene blue degradation activity. ZnO synthesized using succinic acid as fuel exhibited hexagonal wurtzite structure of ZnO and showed highest photocatalytic activity to methylene blue degradation under UV and sun light exposure (Nagaraju *et al.*, 2017). Wurtzite-structured ZnO was synthesized using conventional and microwave combustion method with zinc nitrate as oxidizer, citric acid as fuel, and the fuel-to-oxidizer ratio was kept constant at 1. The pH of the combustion mixture was varied from 2, 7, and 9 by

Figure 3.7. Scanning electron micrographs of ZnO nanoparticles at different annealing temperature. (a) 550°C, (b) 600°C, and (c) 650°C. Reproduced from Yuan *et al.* (2016) with copyright permission (Elsevier).

adding 25% NH$_4$OH. The effect of pH on the rate of combustion reaction and decomposition temperature was studied. The ZnO made by conventional combustion method with small particle size showed high photocatalytic activity than microwave combustion method. The highest photodegradation of methylene blue was observed in conventionally combusted ZnO at pH 7 (Bolaghi *et al.*, 2018). The post combustion annealing temperature was also found to alter the morphology of ZnO (Fig. 3.7) (Yuan *et al.*, 2016). The ZnO annealed at 600°C showed highest efficiency toward methylene blue degradation than other ZnO catalysts because of the presence of highly active and exposed {002} facet. Recently, Bolaghi and his coworkers synthesized ZnO and ZnO/reduced graphene oxide (RGO) composite by the SCS method (Bolaghi *et al.*, 2018; Bolaghi *et al.*, 2018; Bolaghi *et al.*, 2019). The synthesized catalysts were explored for the photocatalytic degradation of methylene blue. Honeycomb-like-structured silver-doped ZnO was synthesized using the SCS method using glucose as fuel (Cai *et al.*, 2013). A 5% doping of Ag on ZnO showed an excellent photocatalytic activity toward rhodamine B degradation because of the existence of open and porous nanostructured surface layer (Fig. 3.8).

The SCS method was extensively used to synthesize doped ZnOs (Yogeeswaran *et al.*, 2006). Lattice-substituted Cd- or In-doped ZnO was formed with CdCl$_2$ and InCl$_3$ as dopant precursors, and (ZnO)$_m$In$_2$O$_3$ was formed when indium nitrate was used as the dopant precursor. The catalysts Cd-doped ZnO and In-doped ZnO showed enhanced photo-responses compared to undoped ZnO. The

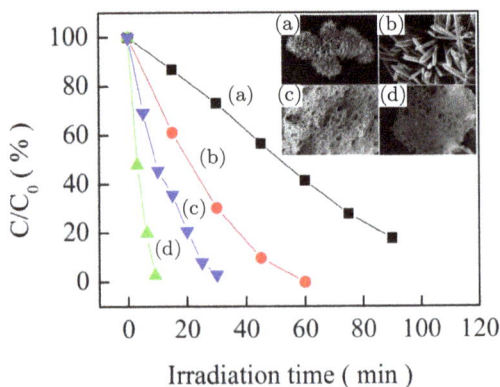

Figure 3.8. Rhodamine B degradation curves for (a) ZnO nanoflowers and (b) ZnO nanorodands via hydrothermal growth, (c) honeycomb pure ZnO, and (d) Ag/ZnO via the SCS method. Insets show the scanning electron microscope (SEM) images of (a)–(d). Reproduced from Cai *et al.* (2013) with copyright permission (Elsevier).

SCS method was used to synthesize the In–Sn co-doped ZnO for the photocatalytic degradation of direct red-31 (DR-31) dye by varying the different aging time (Bhatia *et al.*, 2017). The reaction mixture of zinc nitrate hexahydrate, glucose, and 2% In–Sn co-dopants was aged for 0, 24, and 36 h, and combusted at 400°C. The 24-h aged In–Sn co-doped ZnO nanoparticles showed complete DR-31 dye degradation in 120 min. Rare-earth metal–doped ZnO was also synthesized using the SCS method (Kumar *et al.*, 2015; Bhatia *et al.*, 2017). Ce-doped ZnO and Er-doped ZnO were synthesized and used for the photocatalytic degradation of direct red-23 dye and DR-31, respectively. The optimum concentration of Ce was 3.28% and Er was 2.5% to show the highest degradation efficiency.

3.3.3. *Other Binary Oxides*

Other simple metal oxides were also synthesized using the SCS method and were studied for the different photocatalytic reactions. WO_3 was synthesized using the SCS method with three different fuels, such as glycine, urea, and thiourea (Morales *et al.*, 2008), and the effect of fuel on the band structure was studied in this work. In situ doping of carbon, nitrogen, and sulfur from the fuels showed strong absorption in the visible region, and altered the band gap of WO_3 synthesized using the SCS when compared with commercial WO_3. These catalysts were explored for the photocatalytic degradation of methylene blue under visible light, and the band-engineered WO_3 showed better activity than the commercial P-25 TiO_2. Stoichiometric WO_3 and nonstoichiometric $W_{18}O_{49}$ were synthesized using the SCS method using different fuels such as glycine, urea, and citric acid (Chen *et al.*, 2016). Ammonium paratungstate was used as metal ion salt and ammonium nitrate as oxidizer. Glycine has the combination of both carboxylic acid and amine functional group compared to urea and citric acid. Polymerization of carboxyl group with amine group formed a network structure with cross linking, which produced stoichiometric WO_3. On the other hand, urea and citric acid produced oxide ion vacancy-rich nonstoichiometric $W_{18}O_{49}$ nanoneedles. Optically active metal nanorods of $W_{18}O_{49}$ doped with Fe^{3+} were prepared using the SCS (Chen *et al.*, 2015). The photoluminescence intensity increased with increasing loading of Fe^{3+}, signifying the introduction of defects after Fe^{3+} incorporation. The best efficiency of the catalyst for the photocatalytic degradation of organic compounds was found with 0.5 wt% Fe^{3+}-doped $W_{18}O_{49}$ because of the synergistic effect between catalyst and defects caused by doping. Nanopowders of $W_{18}O_{49}$ with large number oxygen vacancies were synthesized using the SCS using glycine as fuel (Chen *et al.*, 2015). Mixed oxide of WO_3–TiO_2 was also synthesized using the SCS for photocatalytic applications (Singh and Madras, 2013).

Combustion-synthesized copper oxides or copper-doped oxides are another promising semiconductor materials, which are recently being studied for catalysis. Nanoparticles of CuO were synthesized by changing the fuel from glycine to citric acid, and different morphologies were obtained depending on the fuel (Umadevi and Christy, 2013; Christy *et al.*, 2013). The materials showed excellent photocatalytic activity toward methyl orange degradation with pseudo first-order rate constants. Tetragonal rutile-structured Cu-doped SnO_2 quantum dots were synthesized using the SCS approach, and the synthesized materials were employed for the degradation of methyl orange under visible light illumination (Babu *et al.*, 2017). The band gap of SnO_2 decreased with the increase in Cu concentration. The optimum loading of Cu on SnO_2 with highest photocatalytic performance was found to be 0.03%. Further increase in the Cu concentration blocked the surface active sites and generated new centers for recombination of charge carriers on the SnO_2.

Nanocrystalline α-Fe_2O_3 was synthesized using the SCS method using glycine as fuel and ferric nitrate as oxidizer (Jahagirdar *et al.*, 2011). The synthesized material was explored for the photocatalytic degradation of methyl orange. The rate of photodegradation enhanced in the presence of H_2O_2. The influence of fuel-to-oxidizer ratio on the combustion features, phase, morphology, and specific surface area of magnetic α-Fe_2O_3 was evaluated (Cao *et al.*, 2015). The fuel-to-oxidizer ratio was varied from 0.1, 0.2, 0.3, 0.6, 1.0, and 1.2, and the volume of the gases evolved during combustion increased with the increase in fuel ratio. According to propellant chemistry, the maximum energy is released from the reaction, where fuel-to-oxidizer ratio is stoichiometric, and here the stoichiometric ratio between fuel and oxidizer was 0.3, which produced mesoporous hematite with a very high surface area ($103 \, m^2 \, g^{-1}$). The catalyst showed its absorption on the visible light. Iervolino *et al.* (2017) synthesized Ru-doped $LaFeO_3$ by combustion synthesis and investigated for the production of H_2 from the photocatalytic degradation of glucose in both UV and visible light irradiation. The optimum amount of Ru for producing high amount of H_2 was found to be 0.47 mol% (Iervolino *et al.*, 2017).

Combustion-synthesized MoO_2 is another material for photocatalytic exploration, which has been studied thoroughly. Gu *et al.* (2017) synthesized foam-like MoO_2 using the SCS method. Hexaammonium molybdate and glycine with different fuel-to-oxidizer ratio ($\phi = 0.25, 0.5, 0.75, 1.0, 1.25$) were used for the synthesis of MoO_2 nanoparticles (Gu *et al.*, 2017). The combustion temperature was increased when ϕ value increased from 0.25 to 0.5 and then decreased with further increase in ϕ value. The morphology of the synthesized materials varied with different values of ϕ (Fig. 3.9). Foam-like MoO_2 with $\phi = 0.5$ had average particle size of 20–30 nm that showed excellent photocatalytic activity toward the reduction of methylene blue,

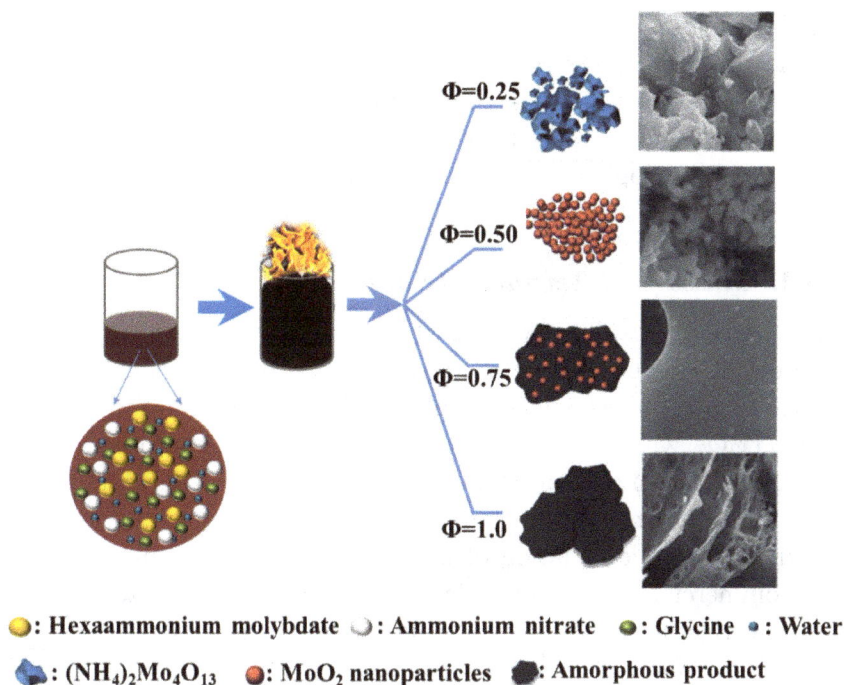

●: **Hexaammonium molybdate** ◯: **Ammonium nitrate** ●: **Glycine** •: **Water**

🐦: **$(NH_4)_2Mo_4O_{13}$** ●: **MoO_2 nanoparticles** ⬛: **Amorphous product**

Figure 3.9. The SCS process for MoO_2 with different morphologies with various Φ values.

methyl orange, rhodamine B, and phenol under visible irradiation. Ultra-porous MoO_3 nanoparticles were synthesized using the SCS (Nagabhushana *et al.*, 2014), and interestingly, for the first time the metal powder was used as precursor, and the synthesized nanoparticles were investigated for the photodegradation of methylene blue. The combustion-synthesized catalyst showed higher activity than commercial MoO_3. CeO_2 is a material of choice for supports in catalysis and have been widely synthesized using the SCS method. CeO_2 from ceric ammonium nitrate and ethylenediaminetetraacetic acid (EDTA) fuel has been studied for photocatalytic reduction of Cr(VI) to Cr(III) or dye degradation (Ravishankar *et al.*, 2015). Umar *et al.* (2015) synthesized nanoflakes of CeO_2 by the SCS method, and they are used for the dye degradation and sensor applications. Combustion-synthesized CeO_2 was also used to evaluate the degradation of methylene blue under irradiation of visible light (Bakkiyaraj and Balakrishnan, 2017).

The experimental setup, the reaction conditions, and the findings over different metal oxides synthesized using the SCS method for the photocatalytic application are tabulated in Table 3.2.

3.3.4. *Mixed and Ternary Oxides*

The composite oxides synthesized using the SCS have also taken a forefront of photocatalytic research. The binary compounds of Bi-oxides are one classic example of this category. Nanoparticles of $Bi_2Ti_2O_7$ with pyrochlore structure synthesized using the SCS showed alteration of band gap on incorporation of Fe and Mn (Samu *et al.*, 2017). The obtained materials showed higher visible light activity for the degradation of methyl orange when compared with the commercial P-25 TiO_2. Gao *et al.* (2016) synthesized BiOCl doped with Eu^{3+} using the SCS method, and they are used for the photodegradation of rhodamine B (Gao *et al.*, 2016). Eu^{3+} doped in the lattice position of BiOCl created oxygen vacancies, and the catalyst showed higher catalytic activity than pristine BiOCl. The Bi-doped BiOCl synthesized using the SCS also showed high activity of photodegradation of rhodamine B under visible light irradiation (Gao *et al.*, 2015). Lv *et al.* (2016) synthesized $NiFe_2O_4/BiOBr$ magnetic composite by the SCS method. The change in the molar ratio of the composite altered the photocatalytic activity toward rhodamine B degradation under visible light irradiation, and the efficient separation of charge with molar ratio of 1:0.2 in $NiFe_2O_4/BiOBr$ was found to be best for the photocatalytic activity. The same group also synthesized $BiVO_4/BiOCl$ heterojunction photocatalyst by the SCS method and studied the visible light degradation of rhodamine B (Lv *et al.*, 2017). $BiVO_4/BiOCl$ composite showed enhanced activity when compared with pristine $BiVO_4$ and BiOCl. The formed heterojunction separated the electrons and holes, and the catalyst exhibited six-cycle photo stability. Monoclinic $BiVO_4$ crystallites were synthesized using the SCS and solid-state reaction method (Jiang *et al.*, 2008). Combustion-synthesized $BiVO_4$ showed

Table 3.2. Screening of solution combustion-synthesized binary oxides in photocatalytic applications, the reaction conditions, and inference.

Catalyst	Synthesis parameters	Reaction conditions	Inference	Reference
WO_3 and $W_{18}O_{49}$	Ammonium paratungstate as W source, glycine, urea, and citric acid	350-W Xe lamp, 100 mL of 40 ppm methylene blue, and 100 mg of catalyst	Considerable rate for organic degradation	Chen *et al.* (2016)
CuO	Copper nitrate as Cu source, citric acid as fuel, and combusted at 300°C	UV light, 0.1 mM of methyl orange, and 0.0125 mM nano CuO	Rate constant: 0.015 min^{-1} for methyl orange degradation	Christy *et al.* (2013)
Cu-doped SnO_2	Tin chloride pentahydrate as Sn source, copper nitrate as Cu source, urea as fuel and combusted at 450°C	200-W Xe lamp, 100 mL of 20 ppm methyl orange, and 100 mg of catalyst	Rate constant: 0.01308 min^{-1} (0.03 mol% Cu/SnO_2), nine times higher than pure SnO_2	Babu *et al.* (2017)
MoO_2	Hexaammonium molybdate as Mo precursor, glycine as fuel	300-W Xe lamp, 50 mL of 40 ppm methyl orange, methylene blue, and rhodamine B, and 100 mg of catalyst	Rate constant: 0.06763 min^{-1} (rhodamine B), 0.10658 min^{-1} (phenol), 0.10567 min^{-1} (methyl orange), and 0.08269 min^{-1} (methylene blue)	Gu *et al.* (2017)

(*Continued*)

Table 3.2.   (*Continued*)

Catalyst	Synthesis parameters	Reaction conditions	Inference	Reference
α-MoO$_3$	Peroxopolymolybdic acid as Mo precursor, sucrose as fuel, and combusted at $470 \pm 10°$C	120-W high-pressure mercury vapor lamp, 250 mL of a 75 ppm methylene blue, and 100 mg catalyst	Combustion MoO$_3$ degraded 100% in 60 min. Commercial MoO$_3$ degraded 18% in 60 min	Nagabhushana *et al.* (2014)
CeO$_2$	Ceric ammonium nitrate as Ce source, ethylenediaminetetraacetic acid as fuel	UV light, 100 mL of 2 ppm trypan blue, and 200 mg of catalyst	CeO$_2$ nanoparticles showed higher percent degradation when compared with bulk CeO$_2$	Ravishankar *et al.* (2015)
α-Fe$_2$O$_3$	Ferric nitrate as Fe source, glycine as fuel, and combusted at $300°$C	125-W high-pressure mercury vapor lamp, 100 mL of a 20 ppm methyl orange, and 100 mg catalyst	100% methyl orange degradation in 25 min	Jahagirdar *et al.* (2011)

spherical morphology for all fuel-to-oxidizer ratio, whereas irregular-shaped BiVO$_4$ was formed in solid-state reaction. The increase in the fuel-to-oxidizer ratio increased the degree of crystallinity. The high crystallinity and less number of defects of the combustion-synthesized BiVO$_4$ crystallite showed 3.5 times higher activity than BiVO$_4$-synthesized using solid-state reaction method. García Pérez *et al.* (2011) synthesized nanospheres of BiVO$_4$ by the SCS method. Sodium carboxymethylcellulose was used as fuel, which also prevented the process of crystal growth and produced 50- to 200-nm spherical BiVO$_4$ nanoparticles. This material showed high photocatalytic activity toward rhodamine B degradation under solar light irradiation. $BiVO_4$ synthesized using the SCS was also used for the photocatalytic degradation of phenol, and H_2 evolution under solar light illumination (Nagabhushana *et al.*, 2013; Perez *et al.*, 2011). Composite of $V_2O_5/BiVO_4$ photocatalysts was synthesized using the one-step SCS method, and 9 mol% $V_2O_5/BiVO_4$ photocatalyst showed highest photocatalytic activity toward methylene blue degradation when compared with other synthesized materials (Jiang *et al.*, 2009). One more composite material, $CuO/BiVO_4$, synthesized using the SCS was used for the photocatalytic degradation of methylene blue. 2 wt% $CuO/BiVO_4$ showed the highest activity of the photoreduction of methylene blue (Jiang *et al.*, 2009). Du and Wang (2019) synthesized $Bi_5O_7NO_3$ and $Ag/Bi_5O_7NO_3$ by the one-pot SCS method, and investigated the photocatalytic degradation of methyl orange in solar light illumination. The photoluminescence spectra of $Ag/Bi_5O_7NO_3$ composite material showed lower intensity than $Bi_5O_7NO_3$. The incorporation of Ag in $Bi_5O_7NO_3$ suppressed the photo-generated electron-hole recombination, and it was least in 5% $Ag/Bi_5O_7NO_3$. $CuBi_2O_4$, the ternary compound, was synthesized using the SCS with 1:1 stoichiometric ratio between fuel and oxidizer (Hossain *et al.*, 2017). Copper nitrate and bismuth nitrate were used as oxidizer precursors, and urea as fuel. Nanocomposite of $CuO/CuBi_2O_4$ and α-$Bi_2O_3/CuBi_2O_4$ was formed by taking excess of copper and bismuth precursor, respectively, in combustion synthesis. The energy band diagram of these composites is shown in Fig. 3.10. The overlap of the bands of CuO and $CuBi_2O_4$ helped to minimize the electron hole recombination by transferring conduction band electrons from CuO to conduction band of $CuBi_2O_4$ and enhanced the charge separation. But over α-$Bi_2O_3/CuBi_2O_4$ because of the absence of proper band alignment, the rate of recombination was higher. Consequently, the nanocomposite of $CuO/CuBi_2O_4$ showed higher activity toward solar water splitting and photoreduction of CO_2 than nanocomposite of α-$Bi_2O_3/CuBi_2O_4$. Novel hetero-structures of $Cs_2O-Bi_2O_3-ZnO$ were synthesized using the SCS method (Hezam *et al.*, 2018). The optimum loading of Cs_2O was 15 mol% to improve the photocatalytic efficiency. The materials

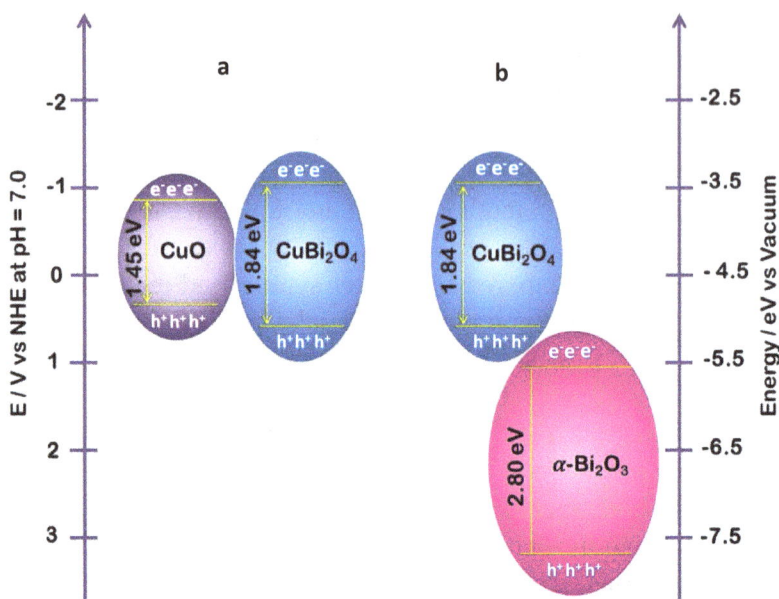

Figure 3.10.   Energy band diagrams for (a) $CuO/CuBi_2O_4$ and (b) α-$Bi_2O_3/$ $CuBi_2O_4$.

were explored for the photocatalytic degradation of 4-chlorophenol. The incorporation of Cs_2O improved the light absorption in both visible and near IR region. The formed direct Z-scheme enhances the activity and stability of the material by decreasing the recombination rate and accumulation of electrons in conduction band and holes in valance band with highest redox potential.

Solution combustion–synthesized spinels were also studied for photocatalytic applications. In an interesting article, the authors have reported the formation of amorphous as well as crystalline $MgAl_2O_4$ spinel nanopowders by the SCS approach (Li *et al.*, 2011). A series of catalysts were prepared by changing the amount of urea during combustion. The increase in the urea content increased the degree of crystallinity and decreased the surface area. The combustion-synthesized catalysts showed good optical absorption in visible region because of the doping of N from the fuel during combustion. Magnetic $CaFe_2O_4$ spinels were synthesized using the SCS and used for the photodegradation of methylene blue under visible light irradiation (Zhang and Wang, 2014). The synthesized material had a particle size of 100 nm and strong optical absorption in the higher wavelength region. The band gap of the $CaFe_2O_4$ nanocatalyst was found to be 1.84 eV. The $CaFe_2O_4$ nanocatalyst completely degraded methylene blue within 45 min, whereas solid-state reaction-synthesized material showed only 25.9% of degradation within the same period of time. The highest photocatalytic activity of $CaFe_2O_4$ nanocatalyst was because of the small particle size. The catalyst was easily recovered using magnet and was recycled. Ag_2WO_4, $CuWO_4$, and $ZnWO_4$ nanoparticles were synthesized using the SCS method, where silver nitrate, copper nitrate, and zinc nitrate were the precursors for Ag, Cu, and Zn, respectively (Thomas *et al.*, 2015). Two different tungsten precursors, sodium tungstate ($Na_2WO_4.2H_2O$) and ammonium tungstate ($(NH_4)_2WO_4$), were used and urea was used as fuel. The combustion mixture was heated in a preheated muffle furnace at 350°C for 5 min. The pure monophasic material formed with sodium tungstate as precursor, whereas biphasic tungstate material formed when ammonium tungstate as precursor. The valance band edge and surface chemical properties of $ZnWO_4$ showed higher photocatalytic activity toward methyl orange degradation than other binary tungstate materials. The perovskite $SrTiO_3$ was synthesized using the combustion of glycine-nitrate-based mixture (Saito *et al.*, 2015). The effect of fuel-to-oxidizer ratio and washing with HNO_3 on the size, crystallinity, and morphology were investigated. The solid product obtained from combustion was calcined at 1000°C for 10 h. The increase in the fuel-to-oxidizer ratio increased the volume of gases emitted after combustion, thus porous-structured compound was obtained. After washing with HNO_3, this porous compound formed highly crystalline particles, which showed good photocatalytic degradation of methylene blue. Another perovskite material $LaFeO_3$ was prepared using the SCS approach (Iervolino *et al.*, 2016). Nitrates of lanthanum and iron were used as oxidizers, and citric acid was used as organic fuel. Set of samples were prepared by varying the amount of citric acid, and the obtained catalysts were probed for the production of H_2 from the photocatalytic degradation of glucose. Parida *et al.* (2010) synthesized orthorhombic

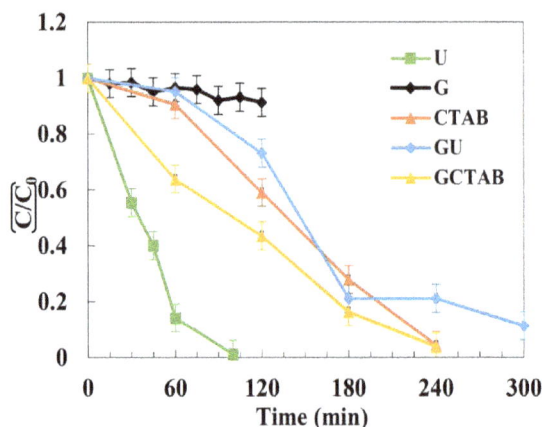

Figure 3.11. Methylene blue degradation curves over the SCS-synthesized BiFeO$_3$ powders using mixture of fuels. (U-urea, G-glycine, CTAB.) Reproduced from Asefi *et al.* (2019) with copyright permission (Elsevier).

LaFeO$_3$ nanocatalyst by sol-gel auto-combustion method and is used for the photodecomposition of water (Parida *et al.*, 2010). The very high amount of H$_2$ was produced with LaFeO$_3$ calcined at 500°C. The calculated quantum efficiency for this catalysts and visible light irradiation was found to be 8.07%. The role of mixed fuels on the synthesis of BiFeO$_3$ by the SCS method was studied recently (Asefi *et al.*, 2019). The combustion with mixed fuels formed impurities such as Bi$_2$Fe$_4$O$_9$ and Bi$_{24}$Fe$_2$O$_{39}$ phases along with BiFeO$_3$, whereas the combustion with urea alone formed pure-phase BiFeO$_3$. During combustion with mixed fuels, the formed intermediates Bi$_2$O$_3$ and Fe$_2$O$_3$ segregated and led to the formation of impurity phases, Bi$_2$Fe$_4$O$_9$ and Bi$_{24}$Fe$_2$O$_{39}$ and these phases were eliminated by post calcination at 600°C. The synthesized materials were used for the photocatalytic degradation of methylene blue. Because of the presence of high purity and crystallinity, BiFeO$_3$ synthesized using urea showed higher rate of photodegradation of methylene blue when compared with mixed fuel combusted BiFeO$_3$ (Fig. 3.11).

The two polymorphs of monoclinic and tetragonal CdWO$_4$ were synthesized using the SCS approach (Eranjaneya and Chandrappa, 2018). The tetragonal CdWO$_4$ showed higher photocatalytic activity than monoclinic phase. The particle size, surface active sites, and life time of charge carriers influence the activity of catalyst. Ternary niobium oxides p-$CuNb_2O_6$ and n-$ZnNb_2O_6$ were synthesized using the SCS method and were used for the photoelectrochemical applications (Kormányos *et al.*, 2016). These materials exhibited high surface area and high crystallinity in nature. Finally, p-$CuNb_2O_6$ showed high efficiency for the photoelectrochemical reduction of CO_2, whereas n-$ZnNb_2O_6$ was active for the photooxidation of sulfite and water. Cubic and orthorhombic phases of Cd_2SnO_4 nanoparticles were produced by the one-step SCS method (Kelkar *et al.*, 2012), and the synthesized materials were explored for the photochemical water splitting under sunlight. The fabricated Cd_2SnO_4 films acted as photoanode.

3.4. Conclusions and Outlook

Apparently, a vast and wide range of photocatalytic materials, such as metal oxides, doped metal oxides, and supported metal oxides, has been synthesized using the solution combustion route and extensively been explored for photocatalytic applications. It is evident from the comprehensive comparison in the literature that primary emphasis has been given over choice of fuel, oxidizer, and fuel-to-oxidizer ratio. These variables during combustion controlled the bulk structure, preferential formation of one polymorph over other, surface morphology, crystallinity, particle size, and so on. The most promising advantage of the SCS is engineering the band gap of the semiconducting material to make it more visible light responsive. Eventually, these properties enhance the photocatalytic efficiency of the materials under visible light exposure. Other than C and N doping, the metal doping also helped the visible light activity of the materials. However, there are contradicting reports on compromising the other physical properties of the catalyst.

In future, it would be really interesting to see the synthesis of more rare-earth-doped materials by the SCS. The rare earths because of its electronic levels are capable of two photon excitation, and solution combustion synthesized up converting nano particles could be the interesting area for the generation of photocatalytic research.

References

Adhikari, S., Gupta, R., Surin, A., Kumar, T. S., Chakraborty, S., Sarkar, D., & Madras, G. (2016). Visible light assisted improved photocatalytic activity of combustion synthesized spongy-ZnO towards dye degradation and bacterial inactivation. *RSC Adv., 6*, 80086–80098.

Asefi, N., Masoudpanah, S. M., & Hasheminiasari, M. (2019). Photocatalytic performances of BiFeO3 powders synthesized by solution combustion method: The role of mixed fuels. *Mater. Chem. Phys., 228*, 168–174.

Assi, N., Mohammadi, A., Sadr Manuchehri, Q., & Walker, R. B. (2015). Synthesis and characterization of ZnO nanoparticle synthesized by a microwave-assisted combustion method and catalytic activity for the removal of ortho-nitrophenol. *Desal. Water Treat., 54*, 1939–1948.

Babu, B., Kadam, A. N., Ravikumar, R. V. S. S. N., & Byon, C. (2017). Enhanced visible light photocatalytic activity of Cu-doped SnO2 quantum dots by solution combustion synthesis. *J. Alloys Compd., 703*, 330–336.

Bakkiyaraj, R., & Balakrishnan, M. (2017). Physical, optical and photochemical properties of CeO2 nanoparticles synthesized by solution combustion method. *J. Adv. Phys., 6*, 41–47.

Bhange, P. D., Awate, S. V., Gholap, R. S., Gokavi, G. S., & Bhange, D. S. (2016). Photocatalytic degradation of methylene blue on Sn-doped titania nanoparticles synthesized by solution combustion route. *Mater. Res. Bull., 76*, 264–272.

Bhatia, S., Verma, N., & Bedi, R. K. (2016). Optical application of Er-doped ZnO nanoparticles for photodegradation of direct red-31 dye. *Optical Mater., 62*, 392–398.

Bhatia, S., Verma, N., & Bedi, R. K. (2017). Effect of aging time on gas sensing properties and photocatalytic efficiency of dye on In–Sn co-doped ZnO nanoparticles. *Mater. Res. Bull., 88*, 14–22.

Bolaghi, Z. K., Hasheminiasari, M., & Masoudpanah, S. M. (2018). Solution combustion synthesis of ZnO powders using mixture of fuels in closed system. *Ceramics Int., 44*, 12684–12690.

Bolaghi, Z. K., Masoudpanah, S. M., & Hasheminiasari, M. (2018). Photocatalytic properties of ZnO powders synthesized by conventional and microwave-assisted solution combustion method. *J. Sol-Gel Sci. Technol., 86*, 711–718.

Bolaghi, Z. K., Masoudpanah, S. M., & Hasheminiasari, M. (2019). Photocatalytic activity of ZnO/RGO composite synthesized by one-pot solution combustion method. *Mater. Res. Bull., 115*, 191–195.

Cai, Y., Fan, H., Xu, M., & Li, Q. (2013). Rapid photocatalytic activity and honeycomb Ag/ZnO heterostructures via solution combustion synthesis. *Colloids Surf. A Physicochem. Eng. Asp., 436*, 787–795.

Cao, Z., Qin, M., Jia, B., Gu, Y., Chen, P., Volinsky, A. A., & Qu, X. (2015). One pot solution combustion synthesis of highly mesoporous hematite for photocatalysis. *Ceramics Int., 41*, 2806–2812.

Challagulla, S., Nagarjuna, R., Ganesan, R., & Roy, S. (2017). TiO2 synthesized by various routes and its role on environmental remediation and alternate energy production. *Nano-Structures & Nano-Objects, 12*, 147–156.

Challagulla, S., & Roy, S. (2017). The role of fuel to oxidizer ratio in solution combustion synthesis of TiO_2 and its influence on photocatalysis. *J. Mater. Res., 32*, 2764–2772.

Challagulla, S., Tarafder, K., Ganesan, R., & Roy, S. (2017). Structure sensitive photocatalytic reduction of nitroarenes over TiO_2. *Sci. Rep., 7*, 8783.

Chen, P., Qin, M., Chen, Z., Jia, B., & Qu, X. (2016). Solution combustion synthesis of nanosized WO_x: Characterization, mechanism and excellent photocatalytic properties. *RSC Adv., 6*(86), 83101–83109.

Chen, P., Qin, M., Liu, Y., Jia, B., Cao, Z., Wan, Q., & Qu, X. (2015). Superior optical properties of $Fe^{3+}–W_{18}O_{49}$ nanoparticles prepared by solution combustion synthesis. *New J. Chem., 39*, 1196–1201.

Chen, P., Qin, M., Zhang, D., Chen, Z., Jia, B., Wan, Q., ... Qu, X. (2015). Combustion synthesis and excellent photocatalytic degradation properties of $W_{18}O_{49}$. *CrystEngComm, 17*, 5889–5894.

Christy, A. J., Nehru, L. C., & Uma devi, M. (2013). A novel combustion method to prepare CuO nanorods and its antimicrobial and photocatalytic activities. *Powder Technol., 235*, 783–786.

Civera, A., Pavese, M., Saracco, G., & Specchia, V. (2003). Combustion synthesis of perovskite-type catalysts for natural gas combustion. *Catal. Today, 83*, 199–211.

Du, X., & Wang, X. (2019). Solution combustion synthesis of Ag-decorated $Bi_5O_7NO_3$ composites with enhanced photocatalytic properties. *Ceramics Int., 45*, 1409–1411.

Eranjaneya, H., & Chandrappa, G. T. (2018). Correction to: Selective synthesis of scheelite/wolframite $CdWO_4$ nanoparticles: A mechanistic investigation of phase formation and property correlation. *J. Sol-Gel Sci. Technol., 85*, 595–595.

Gao, M., Zhang, D., Pu, X., Ding, K., Li, H., Zhang, T., & Ma, H. (2015). Combustion synthesis of Bi/BiOCl composites with enhanced electron–hole separation and excellent visible light photocatalytic properties. *Sep. Purif. Technol., 149*, 288–294.

Gao, M., Zhang, D., Pu, X., Shao, X., Li, H., & Lv, D. (2016). Combustion synthesis and enhancement of BiOCl by doping Eu^{3+} for photodegradation of organic dye. *J. Am. Ceramic Soc., 99*, 881–887.

Gu, S., Qin, M., Zhang, H., Ma, J., Wu, H., & Qu, X. (2017). Facile solution combustion synthesis of MoO_2 nanoparticles as efficient photocatalysts. *CrystEngComm, 19*, 6516–6526.

Hanaor, D. A., & Sorrell, C. C. (2011). Review of the anatase to rutile phase transformation. *J. Mater. Sci., 46*, 855–874.

Hezam, A., Namratha, K., Ponnamma, D., Drmosh, Q. A., Saeed, A. M. N., Cheng, C., & Byrappa, K. (2018). Direct Z-scheme $Cs_2O–Bi_2O_3–ZnO$ heterostructures as efficient sunlight-driven photocatalysts. *ACS Omega, 3*, 12260–12269.

Hossain, M. K., Samu, G. F., Gandha, K., Santhanagopalan, S., Liu, J. P., Jánáky, C., & Rajeshwar, K. (2017). Solution combustion synthesis, characterization, and photocatalytic activity of $CuBi_2O_4$ and its nanocomposites with CuO and α-Bi_2O_3. *J. Phys. Chem. C, 121*, 8252–8261.

Iervolino, G., Vaiano, V., Sannino, D., Rizzo, L., & Ciambelli, P. (2016). Production of hydrogen from glucose by LaFeO$_3$ based photocatalytic process during water treatment. *Int. J. Hydrogen Energy, 41*, 959–966.

Iervolino, G., Vaiano, V., Sannino, D., Rizzo, L., & Palma, V. (2017). Enhanced photocatalytic hydrogen production from glucose aqueous matrices on Ru-doped LaFeO$_3$. *Appl. Catal. B: Environ., 207*, 182–194.

Jahagirdar, A. A., Ahmed, M. Z., Donappa, N., Nagabhushana, H., & Nagabhushana, B. M. (2011). Solution combustion synthesis and photocatalytic activity of α-Fe$_2$O$_3$ nanopowder. *Trans. Ind. Ceram. Soc., 70*, 159–162.

Jiang, H., Nagai, M., & Kobayashi, K. (2009). Enhanced photocatalytic activity for degradation of methylene blue over V$_2$O$_5$/BiVO$_4$ composite. *J. Alloys Compd., 479*, 821–827.

Jiang, H. Q., Endo, H., Natori, H., Nagai, M., & Kobayashi, K. (2008). Fabrication and photoactivities of spherical-shaped BiVO$_4$ photocatalysts through solution combustion synthesis method. *J. Eur. Ceramic Soc., 28*, 2955–2962.

Jiang, H. Q., Endo, H., Natori, H., Nagai, M., & Kobayashi, K. (2009). Fabrication and efficient photocatalytic degradation of methylene blue over CuO/BiVO$_4$ composite under visible-light irradiation. *Mater. Res. Bull., 44*, 700–706.

Kelkar, S. A., Shaikh, P. A., Pachfule, P., & Ogale, S. B. (2012). Nanostructured Cd$_2$SnO$_4$ as an energy harvesting photoanode for solar water splitting. *Energy Environ. Sci., 5*, 5681–5685.

Kim, J., Lee, C. W., & Choi, W. (2010). Platinized WO$_3$ as an environmental photocatalyst that generates OH radicals under visible light. *Environ. Sci. & Technol., 44*, 6849–6854.

Kormányos, A., Thomas, A., Huda, M. N., Sarker, P., Liu, J. P., Poudyal, N., ... Rajeshwar, K. (2016). Solution combustion synthesis, characterization, and photoelectrochemistry of CuNb$_2$O$_6$ and ZnNb$_2$O$_6$ nanoparticles. *J. Phys. Chem. C, 120*, 16024–16034.

Kumar, R., Umar, A., Kumar, G., Akhtar, M. S., Wang, Y., & Kim, S. H. (2015). Ce-doped ZnO nanoparticles for efficient photocatalytic degradation of direct red-23 dye. *Ceramics Int., 41*, 7773–7782.

Kumar, S. G., & Rao, K. K. (2017). Comparison of modification strategies towards enhanced charge carrier separation and photocatalytic degradation activity of metal oxide semiconductors (TiO$_2$, WO$_3$ and ZnO). *Appl. Surf. Sci., 391*, 124–148.

Li, F. T., Ran, J., Jaroniec, M., & Qiao, S. Z. (2015). Solution combustion synthesis of metal oxide nanomaterials for energy storage and conversion. *Nanoscale, 7*, 17590–17610.

Li, F. T., Zhao, Y., Liu, Y., Hao, Y. J., Liu, R. H., & Zhao, D. S. (2011). Solution combustion synthesis and visible light-induced photocatalytic activity of mixed amorphous and crystalline MgAl$_2$O$_4$ nanopowders. *Chem. Eng. J., 173*, 750–759.

Linsebigler, A. L., Lu, G., & Yates Jr, J. T. (1995). Photocatalysis on TiO$_2$ surfaces: Principles, mechanisms, and selected results. *Chem. Rev., 95*, 735–758.

Luttrell, T., Halpegamage, S., Tao, J., Kramer, A., Sutter, E., & Batzill, M. (2014). Why is anatase a better photocatalyst than rutile? Model studies on epitaxial TiO_2 films. *Sci. Rep., 4*, 4043.

Lv, D., Zhang, D., Liu, X., Liu, Z., Hu, L., Pu, X., ... Dou, J. (2016). Magnetic $NiFe_2O_4/BiOBr$ composites: One-pot combustion synthesis and enhanced visible-light photocatalytic properties. *Sep. Purif. Technol., 158*, 302–307.

Lv, D., Zhang, D., Pu, X., Kong, D., Lu, Z., Shao, X., ... Dou, J. (2017). One-pot combustion synthesis of $BiVO_4/BiOCl$ composites with enhanced visible-light photocatalytic properties. *Sep. Purif. Technol., 174*, 97–103.

Ma, X., Xue, L., Li, X., Yang, M., & Yan, Y. (2015). Controlling the crystalline phase of TiO_2 powders obtained by the solution combustion method and their photocatalysis activity. *Ceramics Int., 41*, 11927–11935.

Mani, A. D., Laporte, V., Ghosal, P., & Subrahmanyam, C. (2012). Combustion synthesized TiO_2 for enhanced photocatalytic activity under the direct sunlight-optimization of titanylnitrate synthesis. *Mater. Res. Bull., 47*, 2415–2421.

Morales, W., Cason, M., Aina, O., de Tacconi, N. R., & Rajeshwar, K. (2008). Combustion synthesis and characterization of nanocrystalline WO_3. *J. Am. Chem. Soc., 130*, 6318–6319.

Mukasyan, A. S., Epstein, P., & Dinka, P. (2007). Solution combustion synthesis of nanomaterials. *Proc. Comb. Inst., 31*, 1789–1795.

Nagabhushana, G. P., Nagaraju, G., & Chandrappa, G. T. (2013). Synthesis of bismuth vanadate: Its application in H_2 evolution and sunlight-driven photodegradation. *J. Mater. Chem. A, 1*, 388–394.

Nagabhushana, G. P., Samrat, D., & Chandrappa, G. T. (2014). α-MoO_3 nanoparticles: Solution combustion synthesis, photocatalytic and electrochemical properties. *RSC Adv., 4*, 56784–56790.

Nagaraju, G., Shivaraju, G. C., Banuprakash, G., & Rangappa, D. (2017). Photocatalytic activity of ZnO nanoparticles: Synthesis via solution combustion method. *Mater. Today: Proc., 4*, 11700–11705.

Nagarjuna, R., Challagulla, S., Sahu, P., Roy, S., & Ganesan, R. (2017). Polymerizable sol–gel synthesis of nano-crystalline WO_3 and its photocatalytic Cr (VI) reduction under visible light. *Adv. Powder Technol., 28*, 3265–3273.

Nagaveni, K., Hegde, M. S., & Madras, G. (2004). Structure and photocatalytic activity of $Ti_{1-x}M_xO_{2\pm\delta}$ (M = W, V, Ce, Zr, Fe, and Cu) synthesized by solution combustion method. *J. Phys. Chem. B, 108*, 20204–20212.

Nagaveni, K., Hegde, M. S., Ravishankar, N., Subbanna, G. N., & Madras, G. (2004). Synthesis and structure of nanocrystalline TiO_2 with lower band gap showing high photocatalytic activity. *Langmuir, 20*, 2900–2907.

Nagaveni, K., Sivalingam, G., Hegde, M. S., & Madras, G. (2004a). Photocatalytic degradation of organic compounds over combustion-synthesized nano-TiO_2. *Environ. Sci. & Technol., 38*(5), 1600–1604.

Nagaveni, K., Sivalingam, G., Hegde, M. S., & Madras, G. (2004b). Solar photocatalytic degradation of dyes: High activity of combustion synthesized nano TiO_2. *Appl. Catal. B: Environ., 48*, 83–93.

Nakata, K., & Fujishima, A. (2012). TiO_2 photocatalysis: Design and applications. *J. Photochem. Photobiol. C: Photochem. Rev., 13,* 169–189.

Nakata, K., Ochiai, T., Murakami, T., & Fujishima, A. (2012). Photoenergy conversion with TiO_2 photocatalysis: New materials and recent applications. *Electrochim. Acta, 84,* 103–111.

Ong, C. B., Ng, L. Y., & Mohammad, A. W. (2018). A review of ZnO nanoparticles as solar photocatalysts: Synthesis, mechanisms and applications. *Renew. Sustain. Energy Rev., 81,* 536–551.

Pany, S., Parida, K. M., & Naik, B. (2013). Facile fabrication of mesoporosity driven N–TiO_2@ CS nanocomposites with enhanced visible light photocatalytic activity. *RSC Adv., 3*(15), 4976–4984.

Parida, K. M., Reddy, K. H., Martha, S., Das, D. P., & Biswal, N. (2010). Fabrication of nanocrystalline $LaFeO_3$: An efficient sol–gel auto-combustion assisted visible light responsive photocatalyst for water decomposition. *Int. J. Hydrogen Energy, 35,* 12161–12168.

Pérez, U. G., Sepúlveda-Guzmán, S., Martínez-De La Cruz, A., & Méndez, U. O. (2011). Photocatalytic activity of BiVO4 nanospheres obtained by solution combustion synthesis using sodium carboxymethylcellulose. *J. Mol. Catal. A: Chem., 335,* 169–175.

Potti, P. R., & Srivastava, V. C. (2012). Comparative studies on structural, optical, and textural properties of combustion derived ZnO prepared using various fuels and their photocatalytic activity. *Ind. Eng. Chem. Res., 51,* 7948–7956.

Ramasami, A. K., Naika, H. R., Nagabhushana, H., Ramakrishnappa, T., Balakrishna, G. R., & Nagaraju, G. (2015). Tapioca starch: An efficient fuel in gel-combustion synthesis of photocatalytically and anti-microbially active ZnO nanoparticles. *Mater. Charact., 99,* 266–276.

Ravishankar, T. N., Ramakrishnappa, T., Nagaraju, G., & Rajanaika, H. (2015). Synthesis and characterization of CeO_2 nanoparticles via solution combustion method for photocatalytic and antibacterial activity studies. *Chem. Open, 4,* 146–154.

Rogachev, A. S., & Mukasyan, A. S. (2014). *Combustion for Material Synthesis.* CRC Press.

Roy, S., Aarthi, T., Hegde, M. S., & Madras, G. (2007). Kinetics of photocatalytic reduction of NO by CO with Pd^{2+}-ion-substituted nano-TiO_2. *Ind. Eng. Chem. Res., 46,* 5798–5802.

Roy, S., Hegde, M. S., Ravishankar, N., & Madras, G. (2007). Creation of redox adsorption sites by Pd^{2+} ion substitution in nanoTiO_2 for high photocatalytic activity of CO oxidation, NO reduction, and NO decomposition. *J. Phys. Chem. C, 111,* 8153–8160.

Saito, G., Nakasugi, Y., Sakaguchi, N., Zhu, C., & Akiyama, T. (2015). Glycine–nitrate-based solution-combustion synthesis of $SrTiO_3$. *J. Alloys Compd., 652,* 496–502.

Samu, G. F., Veres, Á., Endrődi, B., Varga, E., Rajeshwar, K., & Janáky, C. (2017). Bandgap-engineered quaternary $M_xBi_{2-x}Ti_2O_7$ (M: Fe, Mn) semiconductor nanoparticles: Solution combustion synthesis, characterization, and photocatalysis. *Appl. Catal. B: Environ., 208*, 148–160.

Selcuk, S., & Selloni, A. (2016). Facet-dependent trapping and dynamics of excess electrons at anatase TiO_2 surfaces and aqueous interfaces. *Nature Mater., 15*, 1107.

Singh, S. A., & Madras, G. (2013). Photocatalytic degradation with combustion synthesized WO_3 and WO_3-TiO_2 mixed oxides under UV and visible light. *Sep., & Purif. Technol., 105*, 79–89.

Sivalingam, G., & Madras, G. (2004). Photocatalytic degradation of poly (bisphenol-A-carbonate) in solution over combustion-synthesized TiO_2: Mechanism and kinetics. *Appl. Catal. A: Gen., 269*(1–2), 81–90.

Sivalingam, G., Priya, M. H., & Madras, G. (2004). Kinetics of the photodegradation of substituted phenols by solution combustion synthesized TiO_2. *Appl. Catal. B: Environ., 51*, 67–76.

Sivaranjani, K., & Gopinath, C. S. (2011). Porosity driven photocatalytic activity of wormhole mesoporous $TiO_{2-x}N_x$ in direct sunlight. *J. Mater. Chem., 21*, 2639–2647.

Sontakke, S., Mohan, C., Modak, J., & Madras, G. (2012). Visible light photocatalytic inactivation of Escherichia coli with combustion synthesized TiO_2. *Chem. Eng. J., 189*, 101–107.

Tasca, J. E., Quincoces, C. E., Lavat, A., Alvarez, A. M., & González, M. G. (2011). Preparation and characterization of $CuFe_2O_4$ bulk catalysts. *Ceramics Int., 37*, 803–812.

Thind, S. S., Wu, G., & Chen, A. (2012). Synthesis of mesoporous nitrogen–tungsten co-doped TiO_2 photocatalysts with high visible light activity. *Appl. Catal. B: Environ., 111*, 38–45.

Thomas, A., Janáky, C., Samu, G. F., Huda, M. N., Sarker, P., Liu, J. P., ... Rajeshwar, K. (2015). Time and energy efficient solution combustion synthesis of binary metal tungstate nanoparticles with enhanced photocatalytic activity. *ChemSusChem, 8*, 1652–1663.

Umadevi, M., & Christy, A. J. (2013). Synthesis, characterization and photocatalytic activity of CuO nanoflowers. *Spectrochim. Acta Part A: Mol. Biomol. Spectrosc., 109*, 133–137.

Umar, A., Kumar, R., Akhtar, M. S., Kumar, G., & Kim, S. H. (2015). Growth and properties of well-crystalline cerium oxide (CeO_2) nanoflakes for environmental and sensor applications. *J. Colloid Interface Sci., 454*, 61–68.

Vasei, H. V., Masoudpanah, S. M., Adeli, M., & Aboutalebi, M. R. (2018). Solution combustion synthesis of ZnO powders using CTAB as fuel. *Ceramics Int., 44*, 7741–7745.

Vasei, H. V., Masoudpanah, S. M., Adeli, M., & Aboutalebi, M. R. (2019a). Photocatalytic properties of solution combustion synthesized ZnO powders using mixture of CTAB and glycine and citric acid fuels. *Adv. Powder Technol., 30*, 284–291.

Vasei, H. V., Masoudpanah, S. M., Adeli, M., & Aboutalebi, M. R. (2019b). Solution combustion synthesis of ZnO powders using various surfactants as fuel. *J. Sol-Gel Sci. Technol., 89*(2), 586–593.

Vasei, H. V., Masoudpanah, S. M., Adeli, M., Aboutalebi, M. R., & Habibollahzadeh, M. (2019). Mesoporous honeycomb-like ZnO as ultraviolet photocatalyst synthesized via solution combustion method. *Mater. Res. Bull., 117*, 72–77.

Vinu, R., & Madras, G. (2008). Synthesis and photoactivity of Pd substituted nano-TiO$_2$. *J. Mol. Catal. A: Chem., 291*, 5–11.

Vinu, R., & Madras, G. (2009). Photocatalytic activity of Ag-substituted and impregnated nano-TiO$_2$. *Appl. Catal. A: Gen., 366*, 130–140.

Wen, W., & Wu, J. M. (2014). Nanomaterials via solution combustion synthesis: A step nearer to controllability. *RSC Adv., 4*, 58090–58100.

Xi, G., Yan, Y., Ma, Q., Li, J., Yang, H., Lu, X., & Wang, C. (2012). Synthesis of multiple shell WO$_3$ hollow spheres by a binary carbonaceous template route and their applications in visible light photocatalysis. *Chem. A Eur. J., 18*, 13949–13953.

Xiao, Q., Si, Z., Yu, Z., & Qiu, G. (2007). Sol–gel auto-combustion synthesis of samarium-doped TiO$_2$ nanoparticles and their photocatalytic activity under visible light irradiation. *Mater. Sci. Eng. B, 137*, 189–194.

Yogeeswaran, G., Chenthamarakshan, C. R., de Tacconi, N. R., & Rajeshwar, K. (2006). Cadmium-and indium-doped zinc oxide by combustion synthesis using dopant chloride precursors. *J. Mater. Res., 21*, 3234–3241.

Yuan, Y., Huang, G. F., Hu, W. Y., Xiong, D. N., & Huang, W. Q. (2016). Tunable synthesis of various ZnO architectural structures with enhanced photocatalytic activities. *Mater. Lett., 175*, 68–71.

Zhang, X., Qin, J., Hao, R., Wang, L., Shen, X., Yu, R., Limpanart, S., Ma, M., & Liu, R. (2015). Carbon-doped ZnO nanostructures: Facile synthesis and visible light photocatalytic applications. *J. Phys. Chem. C*, 119, 20544–20554.

Zhang, X., Qin, J., Xue, Y., Yu, P., Zhang, B., Wang, L., & Liu, R. (2014). Effect of aspect ratio and surface defects on the photocatalytic activity of ZnO nanorods. *Sci. Rep., 4*, 4596.

Zhang, Z., & Wang, W. (2014). Solution combustion synthesis of CaFe$_2$O$_4$ nanocrystal as a magnetically separable photocatalyst. *Mater. Lett., 133*, 212–215.

Chapter 4

Solution Combustion Synthesis for Electrochemistry Applications

Wei Wen[*,‡] and Jin-Ming Wu[†,§]
*College of Mechanical and Electrical Engineering,
Hainan University, Haikou 570228, P.R. China
†State Key Laboratory of Silicon Materials and School of Materials
Science and Engineering, Zhejiang University,
Hangzhou 310027, P.R. China
‡wwen@hainu.edu.cn
§msewjm@zju.edu.cn

4.1. Introduction

The solution combustion synthesis (SCS) is a simple, rapid, and energy-efficient method for the synthesis of materials, which utilizes a self-sustained exothermic reaction from well-mixed reactants. The SCS was first reported by Kingsley and Patil (1988), who synthesized Al_2O_3 using urea and $Al(NO_3)_3 \cdot 9H_2O$. Until now, this simple yet efficient preparation method had widely been used to fabricate various materials, including metal oxides, metals/alloys, metal phosphides, metal phosphates, metal sulfides, and metal silicates (Mukasyan *et al.*, 2007; Aruna and Mukasyan, 2008; Patil *et al.*, 1997, 2002; Li *et al.*, 2015). The SCS, self-propagating high-temperature synthesis (SHS), and flame synthesis are the three subcategories of combustion synthesis (CS) (Birol *et al.*, 2013). However, unlike SHS, which starts from pulverized reactants, the SCS begins from a solution in which the reactants are mixed at a molecular level. Moreover, a large amount of gases is usually released during the combustion process in the SCS, which results in a rapid temperature fall immediately after the reaction and thus hinders the sintering of the products (Varma *et al.*, 2016). Therefore, compared with SHS, the SCS is more suitable for the synthesis of nanomaterials (Wen and Wu, 2014).

Under normal conditions, metal nitrate, fuel, and water are used as oxidant, reductant, and solvent for the SCS, respectively (Patil *et al.*, 2008; Lackner, 2010). For organic fuels, carbon and hydrogen serve as reducing elements (González-Cortés and Imbert, 2013). Certain fuels can form complexes with metal ions to improve the mixing homogeneity of the reactants. For example, urea can serve as a complexing agent to form coordination structures, as shown in Fig. 4.1 (Zhang *et al.*, 2012). Xanthopoulou *et al.* (2018) used nuclear magnetic resonance (NMR) to verify the formation of glycine–nickel nitrate complexes in the solution before combustion. Unlike the other synthetic methods, the driving force for the SCS has been created from its own exothermic chemical energy of the redox reactions (Kim *et al.*, 2011). In other words, the energy input for the SCS is required only to ignite the combustion reaction, rather than to provide continuous energy input.

The phase, microstructure, grain size, and specific surface area of the products using the SCS can be adjusted by the fuel, the fuel-to-oxidizer ratio, the amount of water, the pH value, and the

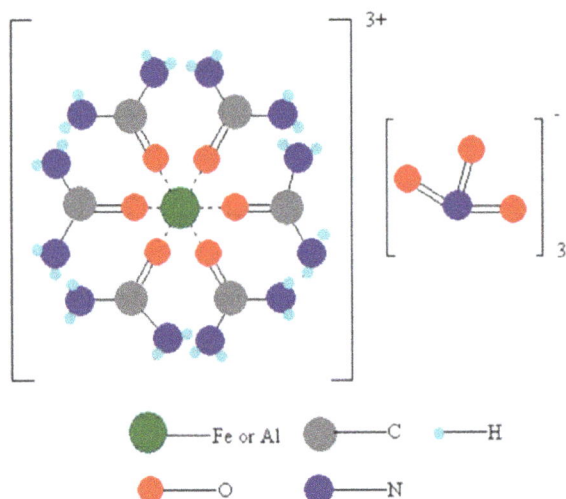

Figure 4.1. Proposed coordination structure of the Fe/Al–urea–nitrate complex (Zhang *et al.*, 2012).

processing parameters (Deganello and Tyagi, 2018). The propellant chemistry theory is widely adopted to calculate the stoichiometry of the fuel and the oxidizer (Jain *et al.*, 1981; Saradhi and Varadaraju, 2006). Taking the combustion reaction between copper nitrate and glycine $(C_2H_5NO_2)$ as an example, the equilibrium combustion reaction can be described as:

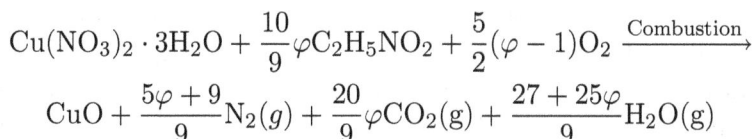

$$Cu(NO_3)_2 \cdot 3H_2O + \frac{10}{9}\varphi C_2H_5NO_2 + \frac{5}{2}(\varphi - 1)O_2 \xrightarrow{\text{Combustion}}$$
$$CuO + \frac{5\varphi + 9}{9}N_2(g) + \frac{20}{9}\varphi CO_2(g) + \frac{27 + 25\varphi}{9}H_2O(g)$$

where $\varphi = 1$ means that atmospheric oxygen is not required for the complete combustion of the fuel, whereas $\varphi < 1$ and $\varphi > 1$ indicate the fuel-lean and fuel-rich conditions, respectively. Considering the synthesis of ZnO as an example, 4.0 g $Zn(NO_3)_2 \cdot 6H_2O$ and 2.42 g urea $(\varphi = 1.8)$ were dissolved in triple-distilled water in a crucible to form a solution, which was then transferred to a preheated furnace maintained at 400°C. After a few minutes, the precursor was ignited to combust and ZnO was obtained (Wen *et al.*, 2013). Because of the high specific surface area and the nanocrystalline nature of the products, the SCS is promising for large-scale fabrications of materials involved in several surface-related procedures like adsorption (Wen and Wu, 2011), catalysis (González-Cortés and Imbert, 2013), and electrochemical energy conversion (Rajeshwar and de Tacconi, 2009) and storage (Wen *et al.*, 2014). This chapter will focus on the SCS for electrochemistry applications.

4.2. Oxygen Reduction Reaction

Electrochemical oxygen reduction reaction (ORR) is an essential reaction that occurs in fuel cells and metal–air batteries (Liu *et al.*, 2018). Platinum (Pt) represents a benchmark for ORR catalysts because of its high activity and good stability, but its applications are hindered by the scarcity and high cost. Thus, developing non-noble metal ORR catalysts with excellent catalytic performances is of great importance. Additionally, economical production of electrocatalysts is of great importance for large-scale industrial applications, which makes the SCS a promising synthetic route (Deganello and Tyagi, 2018).

The ORR performances are usually evaluated using linear sweep voltammetry (LSV) with a rotating disk electrode (RDE). Recently, phosphates were explored as ORR electrocatalysts. A combination of the SCS and calcination can be utilized to synthesize phosphates using $(NH_4)_2HPO_4$ or $NH_4H_2PO_4$ as phosphorus sources. For example, $NaCoPO_4$ and $Na_2CoP_2O_7$ were prepared by Gond *et al.* (2018) using the SCS and a subsequent calcination at 600°C for 5 h. The morphology characterizations are shown in Fig. 4.2, which suggests the formation of homogeneous $NaCoPO_4$ submicrometric particles and $Na_2CoP_2O_7$ sheets (Gond *et al.*, 2018). The $NaCoPO_4$ and $Na_2CoP_2O_7$ can act as effective catalysts for ORR. The half-wave potentials at the $NaCoPO_4$ and $Na_2CoP_2O_7$ are 0.731 and 0.744 V vs. RHE in 1 M NaOH, respectively, which are only slightly lower than that observed at Pt/C (0.804 V) (Gond *et al.*, 2018). Furthermore, the Tafel slopes were found to be 87 and 109 mV decade^{-1} in 1 and 0.1 M NaOH, respectively. Murugesan *et al.* (2018) also explored the applications of sodium and potassium iron phosphates ($AFePO_4$, A = Na and K) synthesized using the SCS as catalysts for ORR. The ORR performances were investigated using cyclic voltammetry (CV)

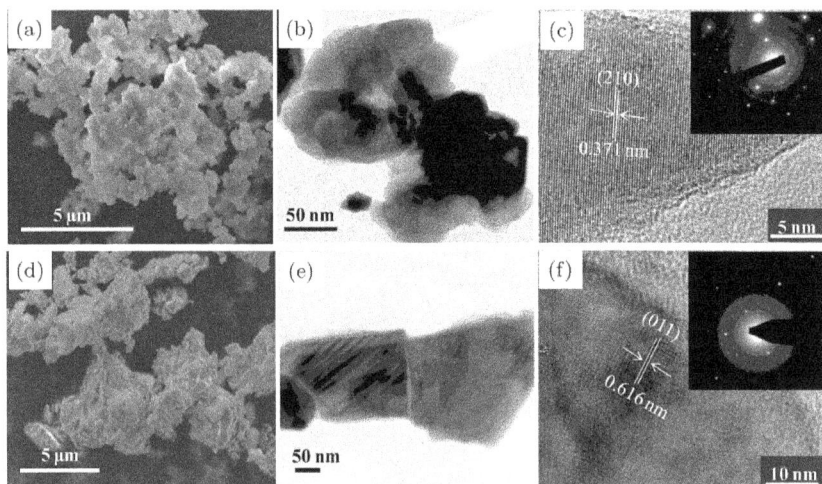

Figure 4.2.   Morphology characterizations of sodium cobalt phosphates. Scanning electron microscopy (SEM), transmission electron microscopy (TEM), and high-resolution transmission electron microscopy (HRTEM) images of $NaCoPO_4$ (a–c) and $Na_2CoP_2O_7$ (d–f). The insets show the corresponding selected area electron diffraction (SAED) pattern (Gond *et al.*, 2018).

and LSV, and the results are shown in Fig. 4.3. The electrocatalytic reduction of oxygen on the $NaFePO_4$ electrode is more efficient, which delivers higher positive onset potential and cathodic current density than that of $KFePO_4$ and $FePO_4$ (Murugesan *et al.*, 2018). Moreover, the $NaFePO_4$ exhibited a high stability over 20 h and an electron transfer number of 3.87 per oxygen molecule, approaching the ideal value of 4 (Murugesan *et al.*, 2018). The half-wave potential at the $NaFePO_4$ was approximately 0.79 V vs. RHE, which is comparable to that of Pt/C (0.86 V) (Murugesan *et al.*, 2018). The excellent ORR performances at the $NaFePO_4$ can be attributed to its orthorhombic crystal structure, the stabilization of active (Fe^{II}) centers using the phosphate framework, and the conductive carbon coating (Murugesan *et al.*, 2018).

Perovskites are promising candidates for electrocatalysis, and interestingly, the SCS is an effective synthesis method for preparing perovskites with low synthetic cost (Wu and Wen, 2010). $LaMO_3$ (M = Cr, Mn, Fe, Co, Ni) perovskites have been synthesized using the SCS with controllable crystallinity, which can be tuned by changing the fuel-to-metal-nitrate ratios (Ashok *et al.*, 2018). The ORR kinetics are in the order of $LaMnO_3 > LaCoO_3 > LaNiO_3 > LaFeO_3 > LaCrO_3$; yet, the stability of $LaCoO_3$ for ORR is better than that of $LaMnO_3$ (Ashok *et al.*, 2018).

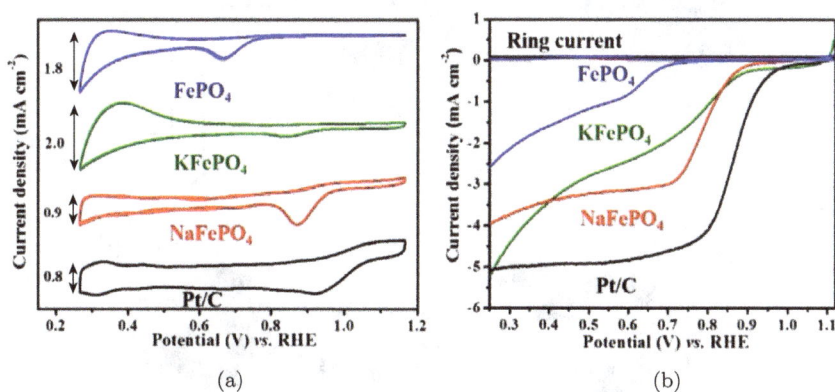

Figure 4.3. Oxygen reduction reaction (ORR) performances of $KFePO_4$ and $NaFePO_4$: (a) Cyclic voltammograms and (b) ORR polarizations. The electrolyte is O_2-saturated 0.1 M KOH. Rotating speed: 1600 rpm, scan rate: 10 mV s^{-1} (Murugesan *et al.*, 2018).

Carbon materials have attracted great interest for electrocatalysis, which fortunately can also be prepared using the SCS combined with a subsequent calcination in inert gas. For instance, hierarchically porous nitrogen-doped carbon was successfully obtained using the SCS, followed by calcination under Ar flow and acid leaching, as shown in Fig. 4.4 (Zhu *et al.*, 2018). Following the SCS and high-temperature (800–1100°C) calcination, an MgO-/nitrogen-doped carbon composite can be obtained, and then the acid leaching removed MgO, achieving the final nitrogen-doped carbon (Fig. 4.4) (Zhu *et al.*, 2018). The achieved nitrogen-doped carbon has a high specific surface area up to $1958\,\mathrm{m^2\,g^{-1}}$ and contains macro-/meso-/micropores (Fig. 4.5). More essentially, the nitrogen-doped carbon shows high ORR activity and good durability, which is among the best reported carbon-based electrocatalysts (Zhu *et al.*, 2018). Combining carbon materials with other electrocatalysts further improves the electrochemical activity. Similar synthetic method can also be used to fabricate Co_9S_8/porous S, N-doped carbon, which has demonstrated a high ORR activity close to that of commercial

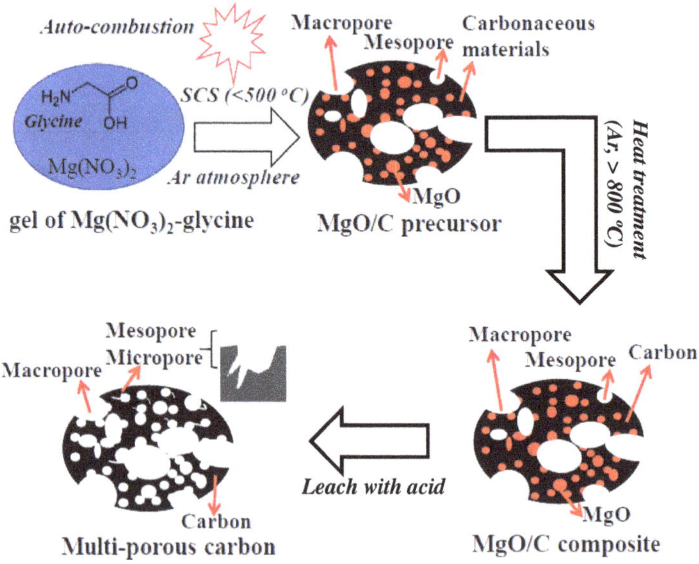

Figure 4.4. Schematics for the formation of the hierarchically porous nitrogen-doped carbon (Zhu *et al.*, 2018).

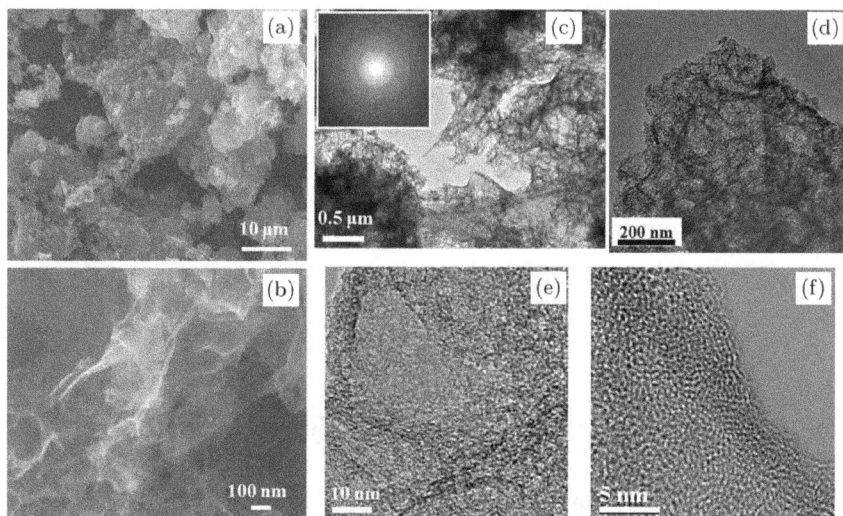

Figure 4.5. Characterizations of the hierarchically porous nitrogen-doped carbon: (a, b) SEM, (c, d) TEM, and (e, f) HRTEM images (Zhu *et al.*, 2018).

Pt/C catalyst, together with superior long-term stability and good tolerance against methanol (Zhu *et al.*, 2017). The catalytic activity of this hybrid catalyst depends on the calcination temperature and glycine amount (Zhu *et al.*, 2017). Similarly, a composite catalyst of Co nanoparticles/nitrogen-doped carbon/carbon nanotubes with high ORR activity was synthesized using a metal nitrate–glycine SCS, followed by high-temperature calcination and acid washing (Zhu *et al.*, 2017).

4.3. Oxygen Evolution Reaction and Hydrogen Evolution Reaction

Electrolytic water splitting to produce hydrogen as a chemical fuel offers a desirable clean energy, owing to the following reasons: high purity of H_2, abundant source, and demanding no high-temperature/high-pressure reactions (Wen *et al.*, 2018; Anantharaj *et al.*, 2016). Water splitting includes two half-cell reactions: hydrogen evolution reaction (HER) at the cathode and oxygen evolution reaction (OER) at the anode. Noble metals such as Pt (cathode) and

Ir and Ru (anode) show high activity for electrolytic water splitting; however, they are scarce and expensive. The parameters used to evaluate catalytic activity for water splitting predominantly include overpotential, Tafel plot, exchange current density, stability, and mass and specific activity (Anantharaj et al., 2016). Electrodes with good conductivity, high surface area, abundant active sites, and excellent wettability are expected to exhibit a high electrocatalytic activity. Compared with HER, OER normally requires a larger overpotential to deliver a desirable current density (e.g., $10\,\mathrm{mA\,cm^{-2}}$), because of the sluggish multistep four-electron process (Wen et al., 2018).

The oxides and hydroxides of Fe, Co, and Ni are widely used for OER. Recently, their sulfides, selenides, and phosphides have also been explored as highly active OER catalysts (Anantharaj et al., 2016). Phosphates can serve as OER catalysts. For example, $NaCoPO_4$ and $Na_2CoP_2O_7$ derived from the SCS exhibited better OER activity than Pt/C, as shown in Fig. 4.6 (Gond et al., 2018). The Tafel slopes are 52 and 51 mV decade^{-1}, respectively. Perovskites also possess attractive performances for OER (Ashok et al., 2018; Xu et al., 2016). It has been reported that, among $LaMO_3$ (M = Cr, Mn, Fe, Co, Ni), $LaCoO_3$ exhibits the highest activity for OER (Ashok et al., 2018).

The SCS has been demonstrated to be capable of synthesizing metastable phases, thanks to the instantaneous high temperature in the combustion process (Deganello and Tyagi, 2018). Furthermore, the SCS diminishes the required calcination temperature and is convenient for achieving cation doping. For example, in $LiNiO_2$, the Ni element in higher oxidation state is expected to show higher OER activity. Al^{3+} doping can stabilize Ni^{3+} in the Ni layer of $LiNiO_2$. Up to 40% of Al could substitute Ni in $LiNiO_2$ using the SCS, whereas only 25% of Al could substitute Ni by solid-state synthesis (Gupta et al., 2015). During high-temperature calcinations, Ni^{3+} is normally reduced to Ni^{2+} by generating oxygen vacancies. Using the SCS, the subsequent crystallization temperature required for $LiNiO_2$ goes below 650°C. The SCS process and the low-annealing temperature are advantageous in keeping the Ni element in its higher oxidation state, resulting in better ordering of Li^+ and Ni^{3+} in their respective

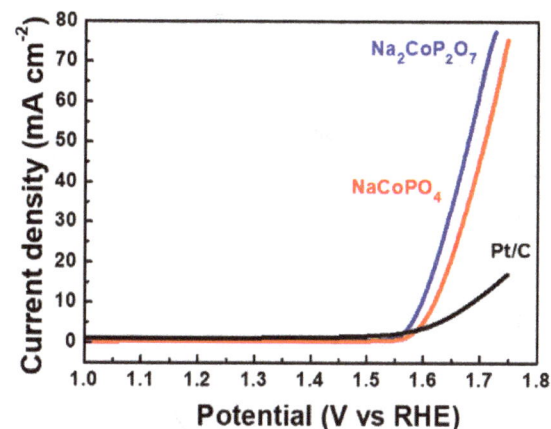

(a)

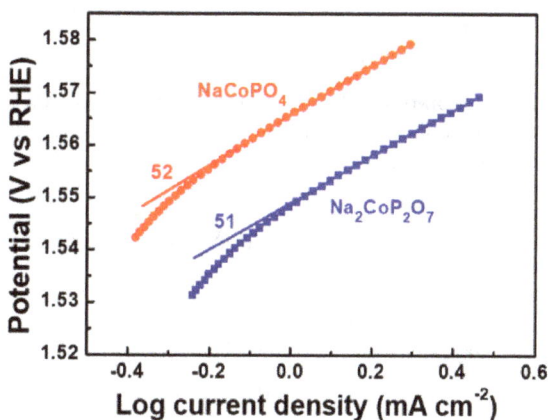

(b)

Figure 4.6. OER performances of $NaCoPO_4$ and $Na_2CoP_2O_7$: (a) OER polarizations and (b) corresponding Tafel plots (Gond *et al.*, 2018).

layers (Gupta *et al.*, 2015). As shown in Fig. 4.7, OER activity of $LiNi_{0.8}Al_{0.2}O_2$ using the SCS is comparable to that of commercial IrO_2 (Gupta *et al.*, 2015). Electrocatalysts with high OER and ORR activities can be used for metal–air batteries. For example, $LaMnO_3$ by the SCS revealed both high ORR and OER catalytic activities and thus could be used for Zn–air batteries with 6 M KOH electrolyte (Zhu *et al.*, 2013). $SmMn_2O_5$ by the SCS could be also utilized as a

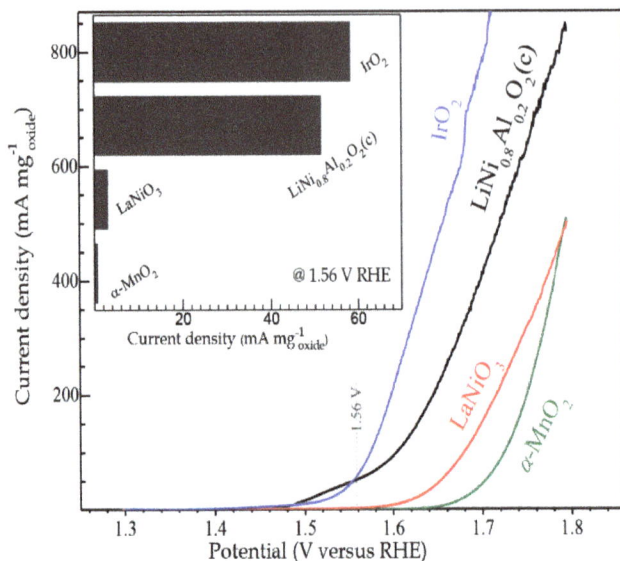

Figure 4.7. Comparison of OER performance of the $LiNi_{0.8}Al_{0.2}O_2$ by the SCS with IrO_2, $LaNiO_3$, and α-MnO_2; the inset shows the current densities at a chosen potential of 1.53 V (Gupta *et al.*, 2015).

cathode for Al–air batteries with 4 M NaOH electrolyte (Chu *et al.*, 2018).

There are relatively only a few reports on using the SCS for HER catalyst synthesis. Metal carbides/carbon composites can be synthesized using the SCS with excess fuel, which is carbonized during the subsequent calcinations, combined with a carbothermal reduction method (Chen *et al.*, 2016; Liu *et al.*, 2018). Chen *et al.* (2016) prepared WC_x/C composite by a combustion–carbothermal reduction method and the synthesis process is shown in Fig. 4.8. During the carbothermal reduction process, WO_3 was converted to WC_x. The obtained WC_x/C composite has a flake-like morphology and the WC_x nanoparticles are embedded in the carbon matrix, as shown in Fig. 4.9 (Chen *et al.*, 2016). The HER on WC_x/C composite exhibits an overpotential of 264 mV to achieve a current density of 10 mA cm^{-2} and the Tafel slope is 85 mV decade^{-1} in 0.5 M H_2SO_4 aqueous solution (Chen *et al.*, 2016). A double core–shell WC–C–Pt can be also fabricated using a similar method, as shown in

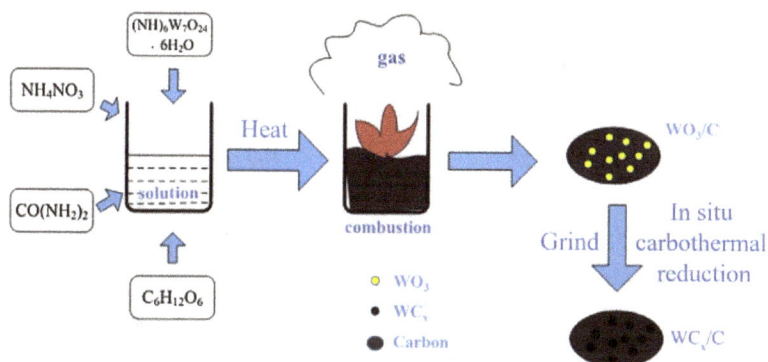

Figure 4.8. Schematics of the synthesis procedure for WC$_x$/C composite (Chen *et al.*, 2016).

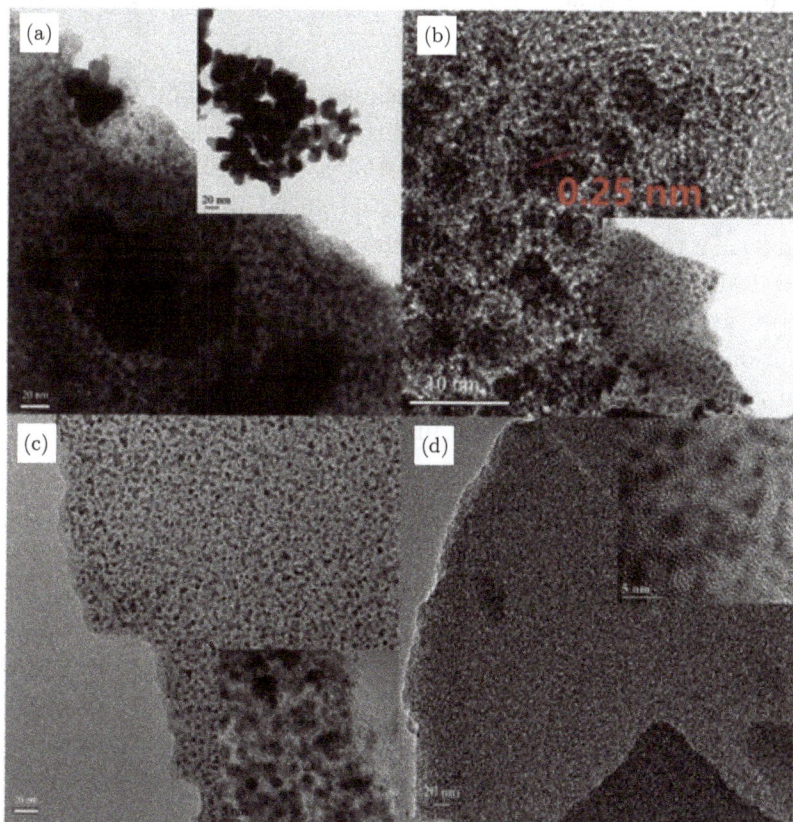

Figure 4.9. TEM images of the WC$_x$/C composites obtained using different glucose/tungsten ratios of (a) 10, (b) 17, (c) 25, and (d) 50 (Chen *et al.*, 2016).

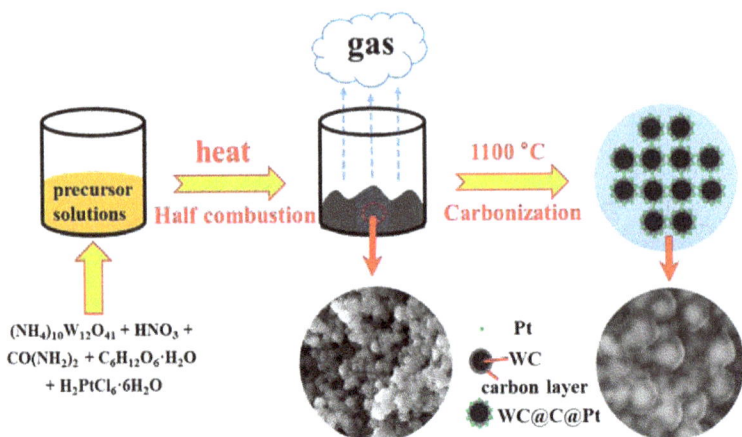

Figure 4.10. Schematics of the synthesis process for WC–C–Pt (Liu *et al.*, 2018).

Fig. 4.10 (Liu *et al.*, 2018). The thickness of the carbon layer and the insertion depth of Pt particles into the carbon layer can be adjusted by changing the carbon content of the precursors (Liu *et al.*, 2018). The achieved WC–C–Pt demonstrated an HER overpotential of 30 and $34\,\text{mV}$ to obtain a current density of $10\,\text{mA cm}^{-2}$ in $0.5\,\text{M H}_2\text{SO}_4$ and $1\,\text{M}$ KOH solution, respectively (Liu *et al.*, 2018). The Tafel slopes are 26 and $27\,\text{mV decade}^{-1}$ in $0.5\,\text{M H}_2\text{SO}_4$ and $1\,\text{M}$ KOH solution, respectively. The excellent electrochemical performances can be attributed to the Pt particles fixed by the carbon layer and good electrical conductivity (Liu *et al.*, 2018).

4.4. Photoelectrochemical Water Splitting

Photoelectrochemical (PEC) water splitting, which directly converts solar energy into chemical fuel hydrogen, offers one of the ideal approaches toward solar energy utilization. The types of PEC water-splitting devices are shown in Fig. 4.11, which can be constructed using a single semiconductor or two semiconductors connected in series (Walter *et al.*, 2010). As shown in Fig. 4.11a, the device is constructed using an *n*-type photoanode (for oxygen evolution) or a *p*-type photocathode (for hydrogen evolution), electrically connected to a hydrogen-evolving or an oxygen-evolving metal electrode. The

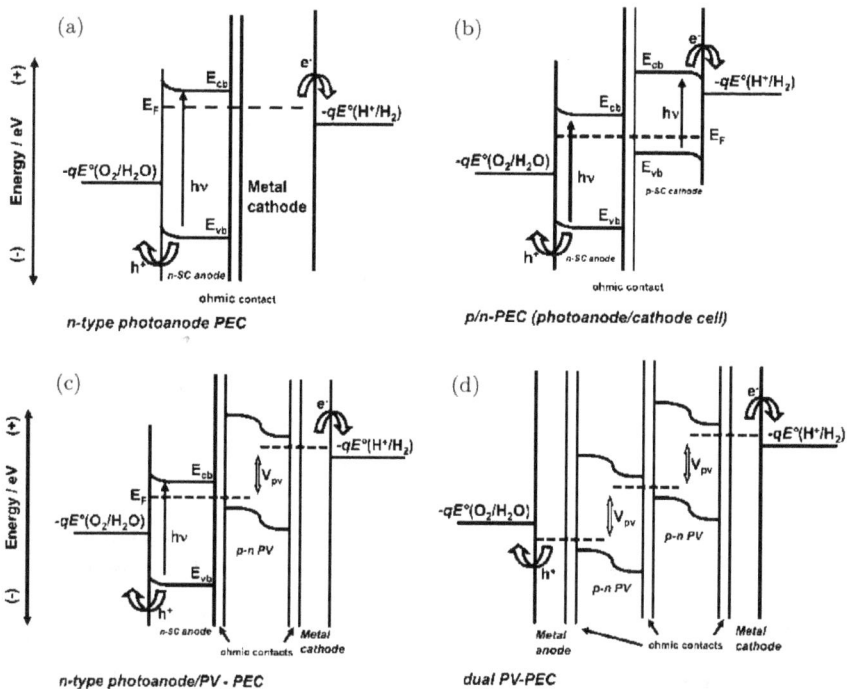

Figure 4.11. Schematics for (a) a single-band gap photoanode PEC device with a metal cathode back contact; (b) a dual-band gap p/n-PEC device with p- and n-type photoelectrodes electrically connected in series; (c) an n-type photoelectrode combined with an integrated p-n PV cell to provide additional bias and connected to a metal cathode; (d) two p-n PV cells connected in series and integrated into metal cathode/anode (Walter *et al.*, 2010).

device in Fig. 4.11b represents a dual-band gap p/n-PEC, which utilizes both minority charge carriers (electron for photocathode and hole for photoanode) to trigger water-splitting reactions at the respective semiconductor/liquid interfaces. In the n-type photoanode/PV device (Fig. 4.11c), holes are generated in the n-type photoanode to oxidize water, and a PV cell is connected in series to provide the additional bias required to reduce protons in solution. In Fig. 4.11d, the device is constructed using two p-n PV cells connected in series and coated with a metal anode and cathode (Walter *et al.*, 2010). The PEC performances can be evaluated in separate electrochemical cells to test individual photoelectrodes (Walter *et al.*,

2010). In contrast to the photocathode, the photoanode has attracted more attention because of its more sluggish water oxidation process.

The band gap and the potentials of conduction band and valence band for semiconductors play an important role on PEC performances. The SCS is convenient for the synthesis of new metal oxides and doping for metal oxides, which can tune the band structure. Kelkar *et al.* (2012) synthesized Cd_2SnO_4 nanoparticles with different phases (cubic and orthorhombic) using the SCS. The size of Cd_2SnO_4 nanoparticles ranges from 10 to 15 nm (Kelkar *et al.*, 2012). The cubic Cd_2SnO_4 shows high activity for PEC (Fig. 4.12), which exhibits a photocurrent of $0.25\,\mathrm{mA\,cm}^{-2}$ at a bias of 0.6 V vs. Ag/AgCl in 1 M NaOH (namely 1.623 V vs. RHE) (Kelkar *et al.*, 2012). Other ternary oxide semiconductors can also be conveniently obtained using the SCS. For example, p-$CuNb_2O_6$ with a band gap of 1.9 eV and n-$ZnNb_2O_6$ with a band gap of 3.2 eV can be fabricated using the SCS followed by a subsequent calcination at 600°C for 30 min, and the morphology characterizations are shown in Fig. 4.13 (Kormányos *et al.*, 2016). The p-$CuNb_2O_6$ revealed attractive activity for the PEC reduction of CO_2 and the n-$ZnNb_2O_6$

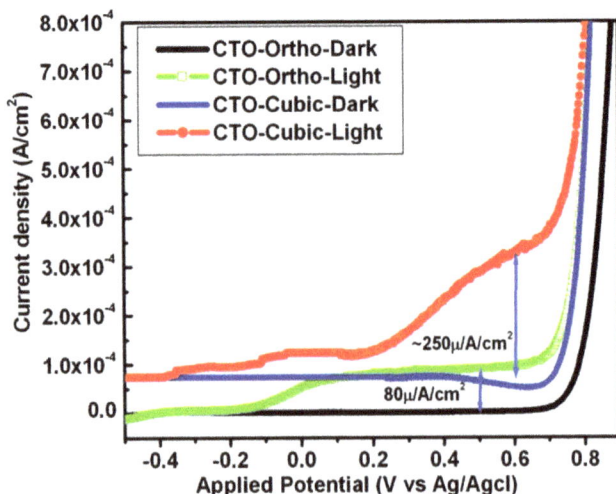

Figure 4.12. Linear sweep voltammetry curves of cubic and orthorhombic Cd_2SnO_4 nanoparticles under light irradiation and dark (Kelkar *et al.*, 2012).

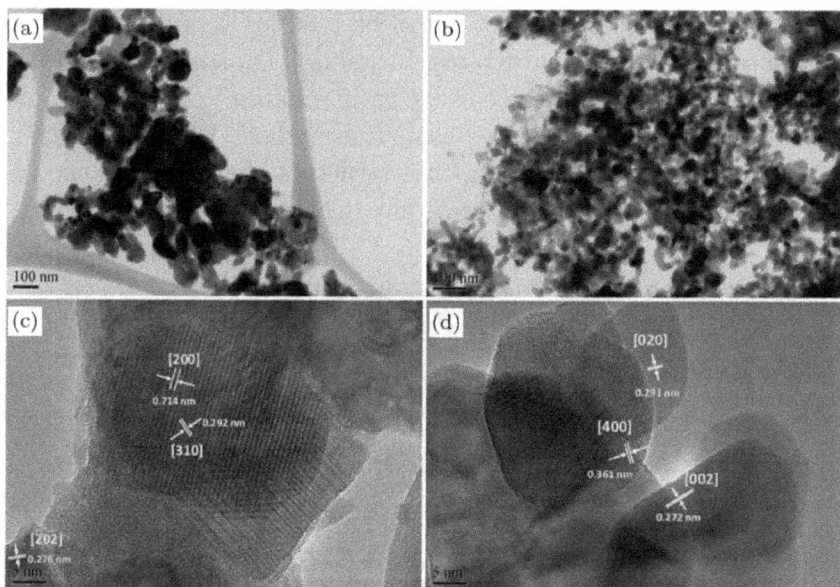

Figure 4.13.   TEM and HRTEM images of $CuNb_2O_6$ (a, c) and $ZnNb_2O_6$ (b, d) calcinated at $600°C$ (Kormányos *et al.*, 2016).

was more suitable for water photooxidation (Kormányos *et al.*, 2016). Varga *et al.* (2018) prepared $CuCrO_2$ with an interconnected porous structure and found that this material can perform PEC carbon dioxide reduction with the products of carbon monoxide, methane, formic acid, methanol, as well as hydrogen generated from water splitting. Moreover, $CuCrO_2$ showed better PEC stability than its monometallic Cu_2O counterpart.

Compared with powders, thin films improve the charge collection efficiency. In particular, single-crystal one-dimensional (1D) arrays not only enhance charge transport, but also decouple the directions of charge-carrier collection and light absorption. However, the PEC performances of 1D arrays are hindered by their insufficient surface area and weak light-harvesting ability. Growing branches can effectively improve surface area of 1D arrays and enhance light-harvesting capability. Wen *et al.* (2017) conducted a room temperature reaction of hydrogen peroxide solution with a Ti-based amorphous complex prepared using the SCS to obtain a precursor solution. This precursor solution can be used for depositing anatase nanosheets on various substrates, including 1D arrays (Wen *et al.*, 2015, 2016, 2017). Balsam-pear-like rutile–anatase core–shell titania nanorod arrays were fabricated using this method (Wen *et al.*, 2017). The morphology, composition, and structure characterizations are shown in Fig. 4.14. The nanoarrays have a balsam-pear-like structure,

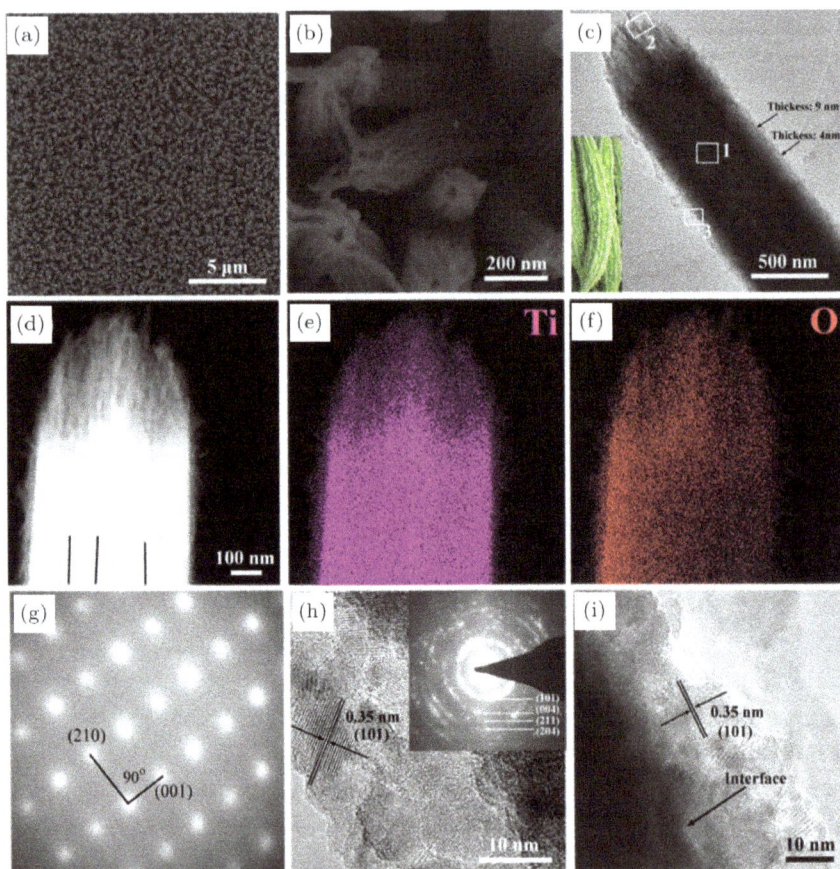

Figure 4.14. Morphology characterizations of rutile/anatase nanorod arrays: (a, b) SEM images, (c) TEM images (inset: a photograph of balsam pears), (d) scanning transmission electron microscopy (STEM) image, (e, f) energy-dispersive X-ray (EDX) mapping, (g) SAED pattern of the trunk, (h) TEM image of the branches on the top of the nanorod (inset: SAED pattern), and (i) TEM image of the branches on the side of the nanorod (Wen *et al.*, 2017).

which is constructed using single-crystal rutile TiO_2 nanorod core and polycrystalline anatase TiO_2 nanosheet branches. The diameters of the core–shell nanorods are 150–650 nm, whereas the thicknesses of the most nanosheet branches are 4–9 nm (Wen *et al.*, 2017). The PEC experiments were performed in a standard three-electrode configuration with the balsam-pear-like rutile–anatase nanorod arrays as a working electrode, saturated calomel electrode (SCE) as a reference electrode, and Pt foil as a counter electrode, respectively. The balsam-pear-like rutile/anatase nanorods demonstrate a much higher photocurrent than rutile nanorods and anatase nanosheets (Fig. 4.15) (Wen *et al.*, 2017). The enhanced PEC performances of the balsam-pear-like rutile/anatase nanorods can be attributed to their multi-advantages: (1) the hierarchical nanostructure is in favor of light scattering and trapping, (2) the high surface area facilitates the reactions at electrode/electrolyte interface, (3) the single-crystal nanorod cores provide direct and rapid paths for majority carrier transfer, and (4) the rutile–anatase junction favors the charge-carrier separation (Wen *et al.*, 2017).

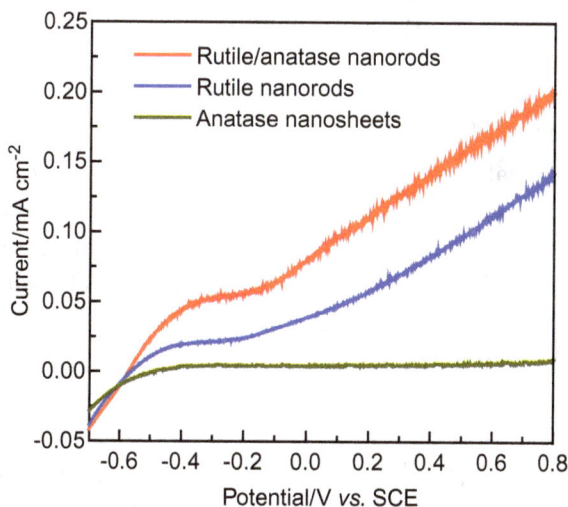

Figure 4.15. Photocurrent curves of rutile/anatase nanorods, rutile nanorods, and anatase nanosheets (Wen *et al.*, 2017).

4.5. Other Electrocatalysis Applications

The SCS-derived nanomaterials have been used for other electro-catalysis applications, such as ethanol oxidation (Matin *et al.*, 2017, 2018). For ethanol oxidation, metal or alloy electrocatalysts are widely used. Fuels in the SCS act as reducing agents; as a result, they can further reduce the metal oxides produced in the SCS using the residual fuel. Up to now, many metals and alloys have been directly synthesized using the SCS (Jung *et al.*, 2005; Rao *et al.*, 2004; Erri *et al.*, 2008; Jiang *et al.*, 2009). The reaction pathways of metal nanopowders in the SCS were investigated in Mukasyan's group both in experiment and thermodynamic analysis (Kumar *et al.*, 2011; Manukyan *et al.*, 2013). The excess reductive gases produced in fuel-rich conditions reduce metal oxides to metals in the reaction front (Manukyan *et al.*, 2013). Matin *et al.* (2018) successfully synthesized Zn-enriched PtZn alloy nanoparticles using the SCS. The PtZn nanoparticles supported on carbon exhibited improved electrocatalytic activity for ethanol oxidation by approximately 2.3 times compared with commercial Pt/C (Matin *et al.*, 2018). Moreover, the amount of fuel in the SCS has great effect on the crystallite size, electronic structure, and electrochemical properties in electrocatalysts (Matin *et al.*, 2018).

4.6. Supercapacitors

Practical applications of many renewable and clean energies, such as solar energy and wind power, heavily rely on advanced energy-storage devices because of their intermittent nature. Supercapacitor (SC) is a promising candidate for electrochemical energy storage. The energy density of SC is much higher than that of an electrolytic capacitor. Although the energy density of SC is lower than that of a battery, it possesses superior charge/discharge ability and cycling lifetime, because the energy storage in SC takes place on the surface/subsurface (Stoller and Ruoff, 2010). Applications of the SCS for lithium ion batteries have been reviewed elsewhere (Wen *et al.*, 2017); in this chapter, only SC applications have been focused on.

SC can be divided into electrochemical double-layer capacitance (EDLC) and pseudocapacitance on the basis of charge-storage mechanism. SC based on EDLC stores and releases energy by nanoscopic charge separation at the electrochemical electrode/electrolyte interface, by primarily employing high-specific area carbon materials (Stoller and Ruoff, 2010). This type of SC possesses rapid charge/discharge capability and long cycle life. Zhu *et al.* (2018) prepared nitrogen-doped carbon with a hierarchically porous structure and a specific surface area of $1958\,\mathrm{m^2\,g^{-1}}$ by a magnesium nitrate–glycine SCS. The effect of calcination temperature and glycine/nitrate ratio on the electrochemical performances was investigated (Zhu *et al.*, 2018). High specific surface area, hierarchically porous structure, heteroatom doping, and excellent electric conductivity are beneficial for the electrochemical performances of carbon materials. High calcination temperature or high glycine/nitrate ratio results in low specific surface area (Zhu *et al.*, 2018). The optimized synthesis conditions for SC are 900°C for calcination temperature and 2 for glycine/nitrate ratio, which lead to a good balance of nitrogen-doping level, graphitization degree, oxygen-doping level, and specific surface area (Zhu *et al.*, 2018). The optimized nitrogen-doped porous carbon shows discharge-specific capacitances of 232, 224, 219, 213, and $205\,\mathrm{F\,g^{-1}}$ at current densities of 1, 2, 3, 5, 7, and $10\,\mathrm{A\,g^{-1}}$, respectively (Zhu *et al.*, 2018).

Compared with EDLC, SC based on pseudocapacitance provides much higher capacitance because of reversible surface faradaic (redox) reactions. Chaturvedi *et al.* (2015) synthesized Li_2MnSiO_4 using the SCS, where urea was used as fuel, and explored its applications as SC. The achieved Li_2MnSiO_4 has a mesoporous structure with a high specific surface area of $35\,\mathrm{m\,g^{-1}}$ (Chaturvedi *et al.*, 2015). Figure 4.16 shows the CV curves from 3 to $50\,\mathrm{mV\,s^{-1}}$ and the rate performance (Chaturvedi et al., 2015). This material exhibits a high specific capacitance of $175\,\mathrm{F\,g^{-1}}$ at $3\,\mathrm{mV\,s^{-1}}$ (Chaturvedi *et al.*, 2015). Srikesh and Nesaraj (2015) obtained NiO nanoparticles using a combination of the SCS with a calcination at 600°C for 3 h and investigated the effect of the type of fuel on the electrochemical performances for SC. The specific capacitance of the NiO nanoparticles derived from glucose is much higher than those prepared using glycine and urea (Srikesh and Nesaraj, 2015). Other metal oxide nanomaterials synthesized using the SCS, such

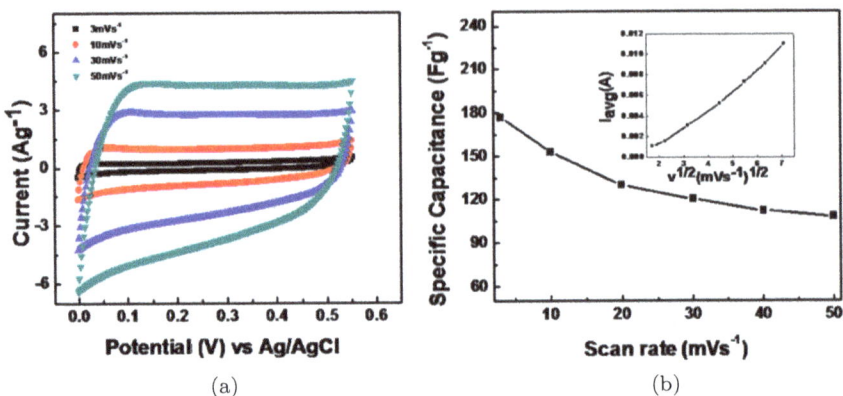

(a)   (b)

Figure 4.16. Electrochemical performances of the Li_2MnSiO_4 electrode: (a) CV curves at different scan rates and (b) rate performance (Chaturvedi *et al.*, 2015).

as Co_3O_4, ZnO, and Fe_2O_3, also revealed high-specific capacitance (Deng *et al.*, 2014; Jayalakshmi *et al.*, 2008; Jayalakshmi and Balasubramanian, 2009).

An appropriate combination of different electrochemically active materials can lead to an expected synergistic effect. The SCS is a convenient route to directly fabricate metal oxide composites. For instance, Zhang *et al.* (2016) successfully obtained $MnO_2/MnCo_2O_4$ composite via a one-step SCS. The amount of the fuel and the Mn/Co ratio have great effects on the phase structure, morphology, and thus electrochemical performances (Zhang *et al.*, 2016). After optimization, the $MnO_2/MnCo_2O_4$ composite showed a high specific capacitance of 497 F g^{-1} at 0.5 A g^{-1} and a relatively good cycling stability at 5 A g^{-1} (Fig. 4.17) (Zhang *et al.*, 2016).

Except for high specific surface area, high electrical conductivity is also required for a high-performance SC, especially for a high-rate performance. The power density of most metal oxides is limited by their low electrical conductivity. In this regard, a combination of metal oxide with carbon nanomaterials with high electrical conductivity, such as graphene, carbon nanotubes, and porous carbon, can effectively tackle this problem. Tao *et al.* (2015) prepared amorphous NiO/C composite through the SCS using citric

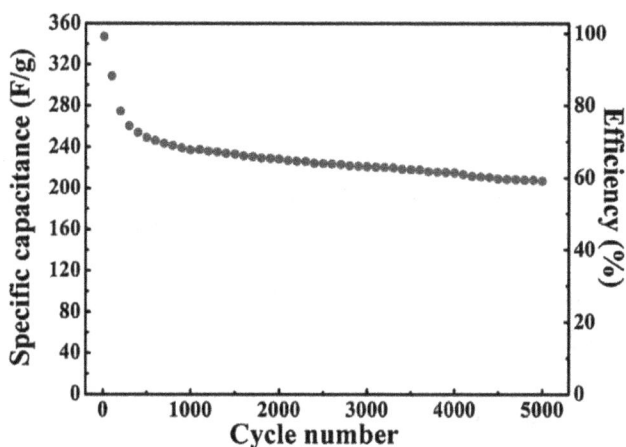

Figure 4.17. Cycling performance of the $MnO_2/MnCo_2O_4$ composite at $5 \, A \, g^{-1}$ (Zhang *et al.*, 2016).

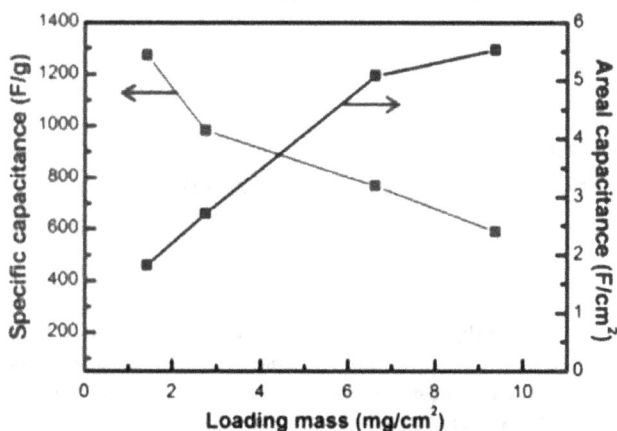

Figure 4.18. Areal capacitance and corresponding specific capacitance of the NiO/C under different mass loading (Tao *et al.*, 2015).

acid as a fuel. The NiO/C composite had a high specific capacitance of $1272 \, F \, g^{-1}$ at $1 \, A \, g^{-1}$ (Tao *et al.*, 2015). When the current density increases to $10 \, A \, g^{-1}$, the specific capacitance is still as high as $642.5 \, F \, g^{-1}$. Furthermore, as the mass loading increases from 1.43 to $9.38 \, mg \, cm^{-2}$, the specific capacitance also retains a high value of $588.5 \, F \, g^{-1}$ (Fig. 4.18) (Tao *et al.*, 2015).

4.7. Conclusion

The SCS is a simple and energy-efficient synthesis method based on a self-sustained combustion reaction, which is suitable for mass productions of materials within a relatively short time. The SCS is attractive for electrochemistry applications, owing to the high specific surface area and the nanocrystalline nature of the obtained products. Moreover, the SCS can be used to synthesize metastable compounds. Up to now, materials derived from the SCS have been widely used in ORR, OER, HER, PEC, electrocatalytic ethanol oxidation, and SCs. Remarkable progresses have been made; however, the exact mechanism of the SCS is still needed to be investigated, which is the prerequisite of fine-tuning the phase compositions and nanostructures of the corresponding electrochemical catalysts/electrodes. Fabrications of high-performance electrocatalysts in thin films or arrays via the SCS also remain a challenge.

Acknowledgments

Dr. Wen is grateful for financial support from the National Natural Science Foundation of China (No. 51862005).

References

Anantharaj, S., Rao Ede, S., Sakthikumar, K., Karthick, K., Mishra, S., & Kundu, S. (2016). Recent trends and perspectives in electrochemical water splitting with an emphasis on sulfide, selenide, and phosphide catalysts of Fe, Co, and Ni: A review. *ACS Catal.*, *6*, 8069–8097.

Aruna, S. T., & Mukasyan, A. S. (2008). Combustion synthesis and nanomaterials. *Curr. Opin. Solid State Mater. Sci.*, *12*, 44–50.

Ashok, A. Kumar, A. Bhosale, R. R. Almomani, F. Malik, S. S. Suslov, S., & Tarlochan, F. (2018). Combustion synthesis of bifunctional $LaMO_3$ ($M = Cr$, Mn, Fe, Co, Ni) perovskites for oxygen reduction and oxygen evolution reaction in alkaline media. *J. Electroanal. Chem.*, *809*, 22–30.

Birol, H., Rambo, C. R., Guiotoku, M., & Hotza, D. (2013). Preparation of ceramic nanoparticles via cellulose-assisted glycine nitrate process: A review. *RSC Adv.*, *3*, 2873–2884.

Chaturvedi, P., Kumar, A., Sil, A., & Sharma, Y. (2015). Cost effective urea combustion derived mesoporous-Li_2MnSiO_4 as a novel material for supercapacitors. *RSC Adv.*, *5*, 25156–25163.

Chen, Z., Qin, M., Chen, P., Jia, B., He, Q., & Qu, X. (2016). Tungsten carbide/carbon composite synthesized by combustion-carbothermal reduction method as electrocatalyst for hydrogen evolution reaction. *Int. J. Hydrog. Energy, 41*, I3005–I3013.

Chu, F., Zuo, C., Tian, Z., Ma, C., Zhao, C., Wang, Y., & Cao, Y. (2018). Solution combustion synthesis of mixed-phase Mn-based oxides nanoparticles and their electrocatalytic performances for Al-air batteries. *J. Alloys Compd., 748*, 375–381.

Deganello, F., & Tyagi, A. K. (2018). Solution combustion synthesis, energy and environment: Best parameters for better materials. *Prog. Cryst. Growth Ch., 64*, 23–61.

Deng, J. C., Kang, L. T., Bai, G. L., Li, Y., Li, P. Y., Liu, X. G., & Liang, W. (2014). Solution combustion synthesis of cobalt oxides (Co_3O_4 and Co_3O_4/CoO) nanoparticles as supercapacitor electrode materials. *Electrochm. Acta, 132*, 127–135.

Erri, P., Nader, J., & Varma, A. (2008). Controlling combustion wave propagation for transition metal/alloy/cermet foam synthesis. *Adv. Mater., 20*, 1243–1245.

Gond, R., Sada, K., Senthikumar, B., & Barpanda, P. (2018). Bifunctional electrocatalytic behavior of sodium cobalt phosphates in alkaline solution. *ChemElectroChem, 5*, 153–158.

González-Cortés, S. L., & Imbert, F. E. (2013). Fundamentals, properties and applications of solid catalysts prepared by solution combustion synthesis (SCS). *Appl. Catal. A-Gen., 452*, 117–131.

Gupta, A., Chemelewski, W. D., Mullins, C. B., & Goodenough, J. B. (2015). High-rate oxygen evolution reaction on Al-doped $LiNiO_2$. *Adv. Mater., 27*, 6063–6067.

Jain, S. R., Adiga, K. C., & Pai Verneker, V. R. (1981). A new approach to thermochemical calculations of condensed fuel-oxidizer mixtures. *Combust. Flame, 40*, 71–79.

Jayalakshmi, M., & Balasubramanian, K. (2009). Solution combustion synthesis of Fe_2O_3/C, Fe_2O_3-SnO_2/C, Fe_2O_3-ZnO/C composites and their electrochemical characterization in non-aqueous electrolyte for supercapacitor application. *Int. J. Electrochem. Sci., 4*, 878–886.

Jayalakshmi, M., Palaniappa, M., & Balasubramanian, K. (2008). Single step solution combustion synthesis of ZnO/carbon composite and its electrochemical characterization for supercapacitor application. *Int. J. Electrochem. Sci., 3*, 96–103.

Jiang, Y., Yang, S., Hua, Z., & Huang, H. (2009). Sol-gel autocombustion synthesis of metals and metal alloys. *Angew. Chem. Int. Ed., 48*, 8529–8531.

Jung, C. H., Jalota, S., & Bhaduri, S. B. (2005). Quantitative effects of fuel on the synthesis of Ni/NiO particles using a microwave-induced solution combustion synthesis in air atmosphere. *Mater. Lett., 59*, 2426–2432.

Kelkar, S. A., Shaikh, P. A., Pachfule, P., & Ogale, S. B. (2012). Nanostructured Cd_2SnO_4 as an energy harvesting photoanode for solar water splitting. *Energy Environ. Sci., 5*, 5681–5685.

Kim, M. G., Kanatzidis, M. G., Facchetti, A., & Marks, T. J. (2011). Low temperature fabrication of high-performance metal oxide thin-film electronics via combustion processing. *Nat. Mater.*, *10*, 382–388.

Kingsley, J. J., & Patil. K. C. (1988). A Novel combustion process for the synthesis of fine particle α-alumina and related oxide materials. *Mater. Lett.*, *6*, 427–432.

Kormányos, A., Thomas, A., Huda, M. N., Sarker, P., Liu, J. P., Poudyal, N., & Rajeshwar, K. (2016). Solution combustion synthesis, characterization, and photoelectrochemistry of $CuNb_2O_6$ and $ZnNb_2O_6$ nanoparticles. *J. Phys. Chem. C*, *120*, 16024–16034.

Kumar, A., Wolf, E. E., & Mukasyan, A. S. (2011). Solution combustion synthesis of metal nanopowders: Nickel-reaction pathways. *AIChE J.*, *57*, 2207–2214.

Lackner, M. (Ed.). (2010). *Combustion Synthesis: Novel Routes to Novel Materials*. Bentham Science Publishers.

Li, F. T., Ran, J., Jaroniec, M., & Qiao, S. Z. (2015). Solution combustion synthesis of metal oxide nanomaterials for energy storage and conversion. *Nanoscale*, *7*, 17590–17610.

Liu, M., Zhao, Z., Duan, X., & Huang, Y. (2019). Nanoscale structure design for high-performance Pt-based ORR catalysts. *Adv. Mater.*, *31*, e1802234.

Liu, Z., Huo, X., Xi, K., Li, P., Yue, L., Huang, M., & Qu, X. (2018). Thickness controllable and mass produced WC–C–Pt hybrid for efficient hydrogen production. *Energy Storage Mater.*, *10*, 268–274.

Manukyan, K. V., Cross, A., Roslyakov, S., Rouvimov, S., Rogachev, A. S., Wolf, E. E., & Mukasyan, A. S. (2013). Solution combustion synthesis of nanocrystalline metallic materials: Mechanistic studies. *J. Phys. Chem. C*, *117*, 24417–24427.

Matin, M. A., Kumar, A., Bhosale, R. R., Saleh Saad, M. A. H., Almomania, F. A., & Al-Marria, M. J. (2017). PdZn nanoparticle electrocatalysts synthesized by solution combustion for methanol oxidation reaction in an alkaline medium. *RSC Adv.*, *7*, 42709–42717.

Matin, M. A., Kumar, A., Saad, M. A. H. S., & Al-Marri, M. J. (2018). Zn-enriched PtZn nanoparticle electrocatalysts synthesized by solution combustion for ethanol oxidation reaction in an alkaline medium. *MRS Commun.*, *8*, 411–419.

Mukasyan, A. S., Epstein, P., & Dinka, P. (2007). Solution combustion synthesis of nanomaterials. *Proc. Combust. Inst.*, *31*, 1789–1795.

Murugesan, C., Lochab, S., Senthilkumar, B., & Barpanda, P. (2018). Earth-abundant alkali iron phosphates ($AFePO_4$) as efficient electrocatalysts for the oxygen reduction reaction in alkaline solution. *ChemCatChem*, *10*, 1122–1127.

Patil, K. C., Aruna, S. T., & Ekambaram, S. (1997). Combustion synthesis. *Curr. Opin. Solid State Mater. Sci.*, *2*, 158–165.

Patil, K. C. Aruna, S. T., & Mimani, T. (2002). Combustion synthesis: An update. *Curr. Opin. Solid State Mater. Sci.*, *6*, 507–512.

Patil, K. C., Hegde, M. S., Yanu, R., & Atuna, S. T. (2008). *Chemistry of Nanocrystalline Oxide Materials: Combustion Synthesis, Properties and Applications*. Singapore: World Scientific.

Rajeshwar, K., & de Tacconi, N. R. (2009). Solution combustion synthesis of oxide semiconductors for solar energy conversion and environmental remediation. *Chem. Soc. Rev.*, *38*, 1984–1998.

Rao, G. R., Mishra, B. G., & Sahu, H. R. (2004). Synthesis of CuO, Cu and CuNi alloy particles by solution combustion using carbohydrazide and N-tertiarybutoxy-carbonylpiperazine fuels. *Mater. Lett.*, *58*, 3523–3527.

Saradhi, M. P., & Varadaraju, U. V. (2006). Photoluminescence studies on Eu^{2+} activated Li_2SrSiO_4 a potential orange-yellow phosphor for solid-state lighting. *Chem. Mater.*, *18*, 5267–5272.

Srikesh, G., & Nesaraj, A. S. (2015). Synthesis and characterization of phase pure NiO nanoparticles via the combustion route using different organic fuels for electrochemical capacitor applications. *J. Electrochem. Sci. Technol.*, *6*, 16–25.

Stoller, M. D., & Ruoff, R. S. (2010). Best practice methods for determining an electrode material's performance for ultracapacitors. *Energy Environ. Sci.*, *3*, 1294–1301.

Tao, K. Y., Li, P. Y., Kang, L. T., Li, X. R., Zhou, Q. F., Dong, L., & Laing, W. (2015). Facile and low-cost combustion-synthesized amorphous mesoporous NiO/carbon as high mass-loading pseudocapacitor materials. *J. Power Sources*, *293*, 23–32.

Varga, A., Samu, G. F., & Janáky, C. (2018). Rapid synthesis of interconnected $CuCrO_2$ nanostructures: A promising electrode material for photoelectrochemical fuel generation. *Electrochim. Acta*, *272*, 22–32.

Varma, A., Mukasyan, A. S., Rogachev, A. S., & Manukyan, K. V. (2016). Solution combustion synthesis of nanoscale materials. *Chem. Rev.*, *116*, 14493–14586.

Walter, M. G., Warren, E. L., McKone, J. R., Boettcher, S. W., Mi, Q., Santori, E. A., & Lewis, N. S. (2010). Solar water splitting cells. *Chem. Rev.*, *110*, 6446–6473.

Wen, W., Chen, Y. N., Wang, S. G., Cao, M. H., Yao, J. C., Gu, Y. J., & Wu, J. M. (2018). A 3D electrode of core–shell branched nanowire $TiN–Ni_{0.27}Co_{2.73}O_4$ arrays for enhanced oxygen evolution reaction. *Appl. Mater. Today*, *12*, 276–282.

Wen, W., & Wu, J. (2011). Eruption combustion synthesis of NiO/Ni nanocomposites with enhanced properties for dye-absorption and lithium storage. *ACS Appl. Mater. Interfaces*, *3*, 4112–4119.

Wen, W., & Wu, J. (2014). Nanomaterials via solution combustion synthesis: A step nearer to controllability. *RSC Adv.*, *4*, 58090–58100.

Wen, W., Wu, J., & Cao, M. (2014). Facile synthesis of a mesoporous Co_3O_4 network for Li-storage via thermal decomposition of an amorphous metal complex. *Nanoscale*, *6*, 12476–12481.

Wen, W., Wu, J., Jiang, Y., Bai, J., & Lai, L. (2016). Titanium dioxide nanotrees for high-capacity lithium-ion microbatteries. *J. Mater. Chem. A*, *4*, 10593–10600.

Wen, W., Wu, J., Jiang, Y., Yu, S., Bai, J., Cao, M., & Cui, J. (2015). Anatase TiO_2 ultrathin nanobelts derived from room-temperature-synthesized titanates for fast and safe lithium storage. *Sci. Rep.*, *5*, 11804.

Wen, W., Wu, J. M., & Wang, Y. D. (2013). Gas-sensing property of a nitrogen-doped zinc oxide fabricated by combustion synthesis. *Sensor. Actuat. B-Chem.*, *184*, 78–84.

Wen, W., Yao, J. C., Gu, Y. J., Sun, T., Tian, H., Zhou, Q. L., & Wu, J. M. (2017). Balsam-pear-like rutile/anatase core/shell titania nanorod arrays for photoelectrochemical water splitting. *Nanotechnology*, *28*, 465602.

Wen, W., Yao, J., Jiang, C., & Wu, J. (2017). Solution combustion synthesis of nanomaterials for lithium storage. *Int. J. Self-Propag. High-Temp. Synth.*, *26*, 187–198.

Wu, J. M., & Wen, W. (2010). Catalyzed degradation of azo dyes under ambient conditions. *Environ. Sci. Technol.*, *44*, 9123–9127.

Xanthopoulou, G., Thoda, O., Roslyakov, S., Steinman, A., Kovalev, D., Levashov, E., & Chroneos, A. (2018). Solution combustion synthesis of nanocatalysts with a hierarchical structure. *J. Catal.*, *364*, 112–124.

Xu, X., Pan, Y., Zhou, W., Chen, Y., Zhang, Z., & Shao, Z. (2016). Toward enhanced oxygen evolution on perovskite oxides synthesized from different approaches: A case study of $Ba_{0.5}Sr_{0.5}Co_{0.8}Fe_{0.2}O_{3-\delta}$. *Electrochim. Acta*, *219*, 553–559.

Zhang, J. S., Guo, Q. J., Liu, Y. Z., & Cheng, Y. (2012). Preparation and characterization of Fe_2O_3/Al_2O_3 using the solution combustion approach for chemical looping combustion. *Ind. Eng. Chem. Res.*, *51*, 12773–12781.

Zhang, Y. Q., Xuan, H. C., Xu, Y. K., Guo, B. J., Lin, H., Kang, L. T., & Du, Y. W. (2016). One-step large scale combustion synthesis mesoporous $MnO_2/MnCo_2O_4$ composite as electrode material for high-performance supercapacitors. *Electrochm. Acta*, *206*, 278–290.

Zhu, C., Aoki, Y., & Habazaki, H. (2017). Co_9S_8 nanoparticles incorporated in hierarchically porous 3D few-layer graphene-like carbon with S, N-doping as superior electrocatalyst for oxygen reduction reaction. *Part. Part. Syst. Charact.*, *34*, 1700296.

Zhu, C., Kim, C., Aoki, Y., & Habazaki, H. (2017). Nitrogen-doped hierarchical porous carbon architecture incorporated with cobalt nanoparticles and carbon nanotubes as efficient electrocatalyst for oxygen reduction reaction. *Adv. Mater. Interfaces*, *4*, 1700583.

Zhu, C., Nobuta, A., Nakatsugawa, I., & Akiyama, T. (2013). Solution combustion synthesis of $LaMO_3$ (M = Fe, Co, Mn) perovskite nanoparticles and the measurement of their electrocatalytic properties for air cathode. *Int. J. Hydrogen Energy*, *38*, 13238–13248.

Zhu, C., Takata, M., Aoki, Y., & Habazaki, H. (2018). Nitrogen-doped porous carbon as-mediated by a facile solution combustion synthesis for supercapacitor and oxygen reduction electrocatalyst. *Chem. Eng. J.*, *350*, 278–289.

Chapter 5

Solution Combustion Approach for the Synthesis of Solid Oxide Fuel Cell Materials and Coatings

S.T. Aruna* and S. Senthil Kumar[†]

Surface Engineering Division, Council of Scientific and Industrial Research-National Aerospace Laboratories, Bangalore 560017, India
* aruna_reddy@nal.res.in
[†] *ssenthil@nal.res.in; sssenkum@gmail.com*

5.1. Fuel Cells

To meet the world's fast-growing energy demand, fossil fuels are being used. Fossil fuels such as gasoline, coal, and jet fuels are nonrenewable. Burning these limited fuel resources not only increases air pollution, but also leads to severe fuel crisis. However, producing power from renewable sources still remains a challenge. Many environmental-friendly alternatives (solar, wind, hydroelectric, and geothermal power) can be used only in particular circumstances (Internet 1). Although batteries play a prominent role in portable devices for storing the energy, they have limited lifetimes and need to be disposed of in hazardous-waste landfills. In contrast, fuel cells are promising energy conversion devices, they can have near-zero emissions, are quiet and efficient, and can work in any environment (Larminie and Dicks, 2003). A fuel cell is an electrochemical conversion device that converts chemical energy into electrical energy with the use of external fuel supply. They are considered to be clean and efficient power-generation devices and some of the merits of fuel cells are as follows: (1) eco-friendly (nonpolluting by-products),

(2) comparatively higher efficiency, (3) reduced weight, especially in mobile applications, (4) fuel flexibility with fast adjustments, (5) low maintenance cost and no moving parts, and (6) long-term stability.

There are different types of fuel cells and are classified based on the type of electrolyte material used in a cell, because it is the property-determining component. There are five main types of fuel cells viz, polymer electrolyte membrane fuel cell (PEMFC), alkaline fuel cells (AFC), phosphoric acid fuel cell (PAFC), molten carbonate fuel cells (MCFC), and solid oxide fuel cells (SOFCs). The electrolyte type, the operating temperature, the charge carrier, power range, and applications of different fuel cells are outlined in Table 5.1. Among all the fuel cells, SOFCs exhibit higher efficiency and has the additional advantage of coupling with gas turbines, thereby increasing the efficiency to ~80%.

5.2. Solid Oxide Fuel Cells

SOFCs represent the second most developed fuel cell technology after PEMFCs. Although PEMFCs are now available in the market for stationary power as a portable consumer electronics option, SOFCs are reaching pre-commercialization as evident from the installation of several hundreds of residential stationary power units (about 1 kW) in Europe and the installation of larger units (250 kW or above) at various utility companies worldwide. SOFCs represent an emerging technology that can offer many advantages, such as high efficiencies, low noise, and reduced emissions over conventional power-generation methods.

In recent years, SOFCs are the most promising and well-explored solution to the emerging pollution-free energy needs of the world. SOFCs are high-temperature fuel cells and are promising because of their higher efficiency and capability to integrate with gas turbines, thereby increasing the efficiency to ~80%. An SOFC is an energy conversion device that combines fuels such as hydrogen, hydrocarbon, and methane with an oxidant like air or oxygen and yields water and electricity. The SOFC is mainly composed of an electrolyte, cathode,

Table 5.1. Different types of fuel cells and their features.

Fuel cell type	PEMFC	AFC	PAFC	MCFC	SOFC
Electrolyte type	Polymer ion exchange membrane	Immobilized alkaline salt solution	Immobilized liquid phosphoric acid	Immobilized liquid molten carbonate	Ionic conducting ceramic oxide
Operating temperature (°C)	30–100	50–200	~220	~650	500–1000
Charge carrier	H^+	OH^-	H^+	CO_3^{2-}	O^{2-}
Power range (W)	1–100 k	500–10 k	10 k–1 M	100 k–10 M	1 k–10 M
Efficiency (%)	60	45	35	45	65–70
Applications	Transport and distributed power generation	Space, transport	Distributed power generation	Combined heat and power system, distributed power generation	Combined heat and power system, distributed power generation
Advantages	High current and power densities, longer operating life	High current and power densities, high efficiency	Advanced technology	High efficiency, internal fuel processing, and high-grade waste heat	Possibility of internal fuel reforming, high-grade waste heat, longer operating life
Limitations	CO intolerance (must be < 20 ppm), water management, and noble metals	CO_2 intolerance	Relatively low efficiency, limited lifetime, use of noble metals	Instability of electrolyte and short operating life	High operating temperature and higher cost

anode, and interconnect. SOFC consists of all solid ceramic parts and operates at high temperatures up to 1000°C. The solid oxide electrolyte coating, which is in contact with a porous anode (air electrode) and cathode (fuel electrode) on either side, is the heart of the fuel cell. The fuel gas is fed to the anode and oxidant gas flows pass the cathode and generates electrical energy by electrochemical oxidation and reduction of the fuel and oxygen, respectively. SOFC operates at very high temperature, ranging from 500°C to 1000°C. Because of all solid-state construction, it is used in wide applications ranging from auxiliary power units (APUs) in vehicles to stationary power generation.

As the operating temperature of the SOFCs is the highest among all fuel cells, it presents both challenges and advantages for the construction of durable SOFC. The high operating temperature (800°C–1000°C) of SOFCs permits rapid kinetics and allows the production of high-quality heat as a by-product suitable for cogeneration. The high-temperature SOFCs have several fundamental advantages over low-temperature fuel cells like PEM, whose operating temperature is <100°C. These advantages include high power density and fuel flexibility. Because of the high operating temperature, it can internally reform the heavier hydrocarbons and electrochemically oxidize both hydrogen and carbon monoxide. Hence, natural gas, which is the most commonly available hydrocarbon fuel, can be utilized in SOFC. Furthermore, it is worth mentioning that other fuel cells like PEM require controlled hydration of the electrolyte membrane and there is an obvious difficulty in maintaining the hydration in colder environments. This problem does not persist in SOFCs as they operate at relatively higher temperatures. Moreover, the heat from spent steam during SOFC operation can be utilized in other process requirements and they can be fed with easy-to-transport fuels instead of pure hydrogen and are highly suited for the use of hydrocarbons for APUs of vehicles as well as for stationary applications. Thus, SOFCs can find application in all types of environments including harsh environments encountered by aircraft, submarines, and so on (Buckingham *et al.*, 2008; Internet 2–5, Gottmann, 2005).

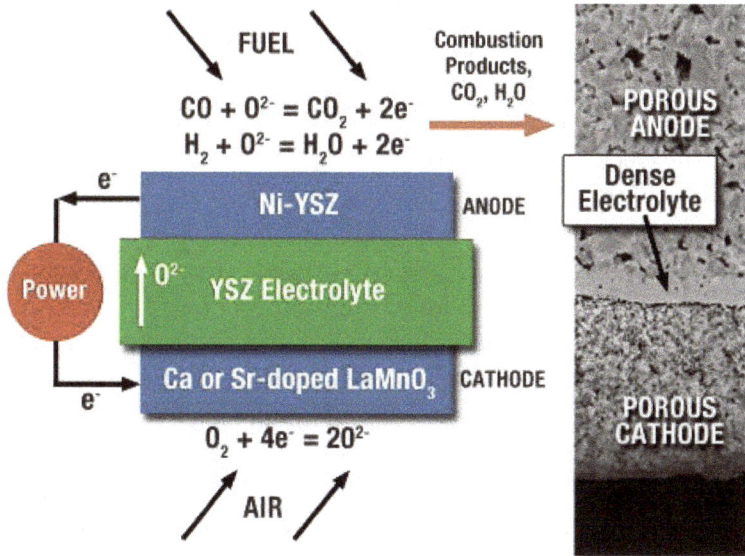

Figure 5.1.   Schematic of single SOFC along with the reactions and microstructure. Adapted from Internet 5.

5.2.1. *Operating Principle of SOFCs*

An SOFC single cell consisting of porous anode and cathode separated using a dense electrolyte membrane is shown in Fig. 5.1. During the operation of SOFC, oxygen (O_2) is fed at the cathode and hydrogen (H_2) as fuel is fed at the anode. The electrolytes are oxygen ion conductors that are permeable only to O^{2-} ions. Molecular oxygen gets reduced on the porous cathode surface by electrons into O^{2-} ions. The oxide ions diffuse through the electrolyte to the fuel-rich and porous anode, in which they react with hydrogen ions (H^+) to produce water as the by-product.

Typical reactions at anode and cathode, and overall reaction using H_2 as fuel in SOFC are shown below:

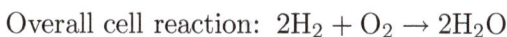

$$\text{Anode reaction: } 2H_2 \rightarrow 4H^+ + 4e^-$$

$$\text{Cathode reaction: } O_2 + 4e^- \rightarrow 2O^{2-}$$

$$\text{Overall cell reaction: } 2H_2 + O_2 \rightarrow 2H_2O$$

The electrochemical oxidation of one H_2 molecule simultaneously gives off two electrons, which flow from anode to cathode through an external circuit as electricity. Power generation is continued as long as fuel and oxidants are supplied. Unlike combustion process, in SOFC, the electrons transferred during oxidation are routed through the external load in the form of electricity.

Fuel cells using hydrogen fuel are much more efficient than internal combustion engines and produce no harmful emissions. But, hydrogen gas also requires a lot of work to produce it from other fuels, thus increasing its production cost. It is difficult to move and store H_2 in gaseous form as high pressure is needed to achieve reasonable storage and it is also exceptionally flammable.

SOFC is a fuel-flexible device and apart from using pure hydrogen, it can utilize carbon monoxide (CO), methane (CH_4), or any other higher hydrocarbon. It can also tolerate some degree of common fossil-fuel impurities, such as ammonia and chlorides.

5.2.2. *SOFC Materials*

The various SOFC materials, the basic requirements of SOFC components along with their fabrication processes, are listed in Table 5.2.

5.2.2.1. *Electrolyte*

The electrolyte is a dense oxide layer that conducts only the oxygen ions and not the electrons (Biswas and Chaitanya, 2013). There are several criteria that the electrolyte has to meet. It should be (1) dense and leak tight without any porosity, (2) stable in reducing and oxidizing environments, (3) a good ionic conductor at operating temperatures, (4) nonelectronic conductor, (5) sufficiently thin to reduce ionic resistance, (6) thermal shock resistance, and (7) cost-effective.

The most popular electrolyte material is 8 mol% Y_2O_3-stabilized ZrO_2 (YSZ). The SOFC operates at 600°C–1000°C in which the ceramic electrolyte becomes conductive to oxygen ions (O^{2-}) but nonconductive to electrons. The addition of yttria to zirconia will

Table 5.2. SOFC components, their requisite properties, and fabrication processes.

Component	Requisite properties	Microstructure	Materials	Fabrication processes
Cathode electrode	High ionic and electronic conductivity; stability in high-temperature oxidizing atmosphere; thermal expansion coefficient matching with other components	Porous	• $LaCoO_3$ • $LaMnO_3$ • $LaSrMnO_3$ • $La_{1-x}Sr_x Co_{1-y}Fe_yO_{3-\delta}$	Plasma spraying, electrochemical vapor deposition, screen printing, sputtering, extrusion
Anode electrode	High electronic conductivity; stability in high-temperature; thermal expansion coefficient matching with other components	Porous	• Ni • $Ni\text{-}ZrO_2$ • $Ni\text{-}Al_2O_3$ • $Ni\text{-}Ce_{1-x} Gd_xO_{2-\delta}$	Tapecasting, plasma spraying, slurry painting, sputtering, pressing
Electrolyte	High ionic conductivity; stable in the high-temperature; Impermeable to gas; thermal expansion coefficient matching with other components	Dense	• $Y_2O_3\text{-}ZrO_2$ • $La_{1-x}\,Sr_x Ga_{1-y} Mg_yO_{3-\delta}$ • $Ce_{1-x}\,Gd_xO_{2-\delta}$	Tapecasting, plasma spraying, slurry painting, sputtering, pressing
Interconnect	High ionic and electronic conductivity; impermeable to gas; stable in high-temperature oxidizing and reducing atmosphere; thermal expansion coefficient matching with other components	Dense	• Ni alloy-Al_2O_3 • Ni alloy-calcia-stabilized zirconia • $LaCrO_3$, • Crofer steel with a protective coating	Plasma spraying, suspension plasma spraying

Table 5.3.   Popular SOFC electrolyte materials and their properties.

Material	Composition	Minimum operating temperature (°C)	TEC $(10^{-6}\mathrm{K}^{-1})$	Conductivity at 750°C (Scm^{-1})
8 mol% yttria-stabilized zirconia (YSZ)	$(Y_2O_3)_{0.08}$ $(ZrO_2)_{0.92}$	700–1000	10.8	1.8×10^{-2}
Gadolinia GDC	$Gd_{0.2}Ce_{0.8}O_{2-\delta}$	550–800	13.5	7.9×10^{-2}
Samaria-doped ceria (SDC)	$Ce_{0.8}$ $Sm_{0.2}O_{1.9}$	550–800	12.2	9.5×10^{-2}
Magnesium-doped lanthanum gallate (LSGM)	$La_{0.9}Sr_{0.1}$ $Ga_{0.8}Mg_{0.2}$ $O_{3-\delta}$	550–800	11.1	9.8×10^{-2}

lead to the formation of compensated oxygen vacancies as per the following defect equation:

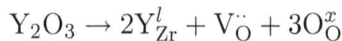

$$Y_2O_3 \rightarrow 2Y_{Zr}^l + V_O^{\cdot\cdot} + 3O_O^x$$

These vacancies act as charge carriers and conduct oxygen ions across the electrolyte. The main disadvantage of YSZ is its high ohmic resistance at low temperature. Numerous other electrolyte materials have been developed to lower the operating temperature of SOFC. Gadolinia-doped ceria (GDC) is one such electrolyte material known for its good ionic conductivity at a relatively lower temperature (550°C–750°C). Some of the common electrolytes are listed in Table 5.3.

5.2.2.2. *Electrodes*

The electrodes (cathodes and anodes) are composed of a reduction/oxidation catalyst that hosts electrochemical reactions (Sun *et al.*, 2010; Mahato *et al.*, 2015; Sun and Stimming, 2007). To ensure a smooth operation of the fuel cell, they have to possess the following properties: (1) high electrical conductivity, (2) high catalytic activity,

(3) high surface area, and (4) compatibility with the electrolyte (and interconnect).

SOFC cathodes are fabricated mostly from electronically conducting oxides or mixed electronically conducting and ion-conducting ceramics. Strontium-doped $LaMnO_3$ is usually applied as the cathode material because it meets the requirements of low cost and it has a high electrical conductivity at high temperatures, and its thermal expansion coefficient matches well with YSZ, a common SOFC electrolyte material. Lanthanum strontium cobalt ferrite (LSCF)-based materials are considered to be candidate cathode materials for intermediate-temperature SOFC (Mahato *et al.*, 2015; Sun and Stimming, 2007; Shri Prakash *et al.*, 2014; Wu and Gupta 2010). LSCF exhibits a high electrical conductivity of $275\,S\,cm^{-1}$ at 600°C (Mahato *et al.*, 2015). Similarly, anodes are made of mixed conducting oxides or electronically conducting metals and ionic conducting ceramics, familiarly known as cermet. The oxide ionic conductivity is vital for SOFC electrodes to support electrochemical reaction by the supply of O^{2-} from the electrolyte. Typically, the anode is made of nickel oxide (NiO)/YSZ, which gets reduced to Ni-YSZ cermet during the operation of SOFC and the cathode is lanthanum-deficient Sr-doped $LaMnO_3$ (Mahato *et al.*, 2015). At higher temperatures, the ionic conductivity of ceramic electrolyte increases and accordingly the electrochemical reactions are enhanced.

As ceramic materials are susceptible to mechanical thermal shock, there are significant restrictions on cell design. The major concerns of SOFC fabrication are thermal expansion mismatch between different ceramic and metallic (interconnect) components and high-temperature sealing difficulties.

5.2.2.3. *Interconnect*

Interconnect in the SOFC not only separates air and fuel gases, but also electrically connects the single cells together (Wu and Gupta, 2010). Mostly, Sr-doped $LaCrO_3$ that exhibits a conductivity of about $15\,ohm^{-1}cm^{-1}$ at 1000°C was used as the interconnect during the initial SOFC development stage (Wu and Gupta, 2010). In recent

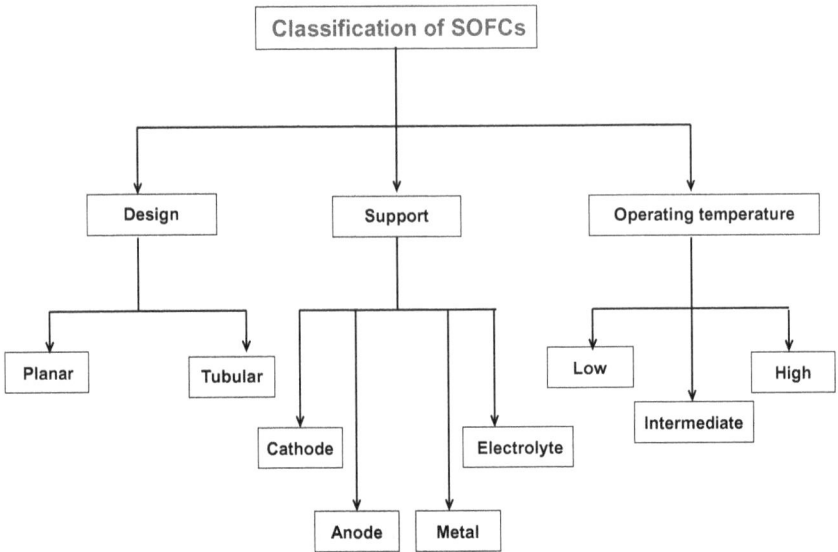

Figure 5.2. Flowchart showing the classification of SOFCs.

years, to reduce the cost of SOFC, Crofer steel is being used as the interconnect plate in both planar SOFC and in metal-supported low-temperature SOFCs (Wu and Gupta, 2010). However, the Crofer steel undergoes oxidation at high temperature and hence lanthanum chromite or Co-Mn spinel oxide is used as a diffusion barrier layer that prevents the diffusion of volatile chromium species from the steel plates into the SOFC layers.

5.2.3. *Design of SOFCs*

SOFCs are classified based on the design, support, and operating temperature (Fig. 5.2) (Singhal and Kendall, 2003).

Though tubular SOFC was the most popular design at the beginning of SOFC development, in the past one decade, planar SOFC design has gained prominence because of its ease of fabrication and higher power density. The differences between the tubular and planar SOFCs are summarized in Table 5.4. In the planar configuration, the electrodes, electrolytes, and the current collectors are present in a flat planar geometry. The interconnection, which is

Table 5.4.   Differences between the tubular and planar SOFCs.

Characteristics	Tubular	Planar
Power density	Low	High
High-temperature sealing	Not necessary	Required, possess problems
Start-up cool down	Faster	Slower
Interconnect	Low cost (a thin strip)	High cost
Manufacturing cost	High	Low
Pressurization	Possible	Not possible
Technology	Redundant	Commercial units

ribbed on both sides, forms gas flow channels and serves as a bipolar gas separator contacting the anode and the cathode of adjoining cells. The planar cells are further subclassified based on the supporting component.

The electrolyte-supported (ES) SOFC design is the oldest design among all and consists of a thick electrolyte (YSZ) layer that offers mechanical support in addition to regular functionality of the electrolyte. Because of the thick electrolyte, these cells have to be operated at very high temperature ($\sim$1000°C). The dense electrolyte layer is important for all fuel cell designs so as to prevent fuel/oxygen crossover. The thick electrolyte has both sides coated with thin electrodes (anode and cathode). The thickness of the electrolyte is the dominant factor that determines the ohmic overpotential and contributes to the performance of the cell. Although ES SOFCs is a more mature configuration, its performance is limited mainly by the high ohmic losses in the thick electrolyte as the oxygen ion crosses from the cathode to the anode, which reduces the power produced. Thus, ES SOFCs show lower efficiency than electrode-supported cells.

The ohmic resistance is much lower in the electrode-supported SOFCs because of the thinner electrolyte ($\sim$5–20 μm) and hence they are better suited for operation at lower temperatures ($\sim$800°C). Because of the lower operating temperature, there will be less degradation of cell and stack components, inexpensive metallic interconnects can be used, there will be less demand on seals, and

SOFCs can be heated up and cooled down at a faster rate. The concentration overpotential is primarily influenced by the electrode thickness. Hence, attention must be paid to the porosity of the supporting electrode to promote gas transport through the electrode.

In the anode-supported SOFCs (ASC) design, the anode cermet layer is the thickest of all components offering the advantage of having smaller ohmic resistance as compared to the ES SOFC design and making them better suited for operation at a lower temperature of 700°C–800°C. Similarly, in cathode-supported cells (CSC), a thick cathode forms the supporting structure (Singhal and Kendall, 2003). Usually, cathodes are oxides like lanthanum strontium manganite, whereas anodes are composed of metal-ceramic composites like Ni-YSZ. Hence, ASC is a preferable design among the planar design because of its low ohmic resistance.

In the tubular design (Fig. 5.3), all the components are assembled in the form of a hollow tube, with the cell constructed around a tubular cathode/anode; air/fuel flows through the inside of the tube (Singhal and Kendall, 2003). The sealing issues faced by the flat design are effectively addressed by tubular design. The long tubular design helps to keep the hermetic sealing portion away from the hot zone and it is possible to adopt a wide range of low-temperature hermetic sealing options. Similar to the planar design, it is also available in electrolyte and electrode-supported configurations. The major disadvantage of tubular design is its poor current collection efficiency.

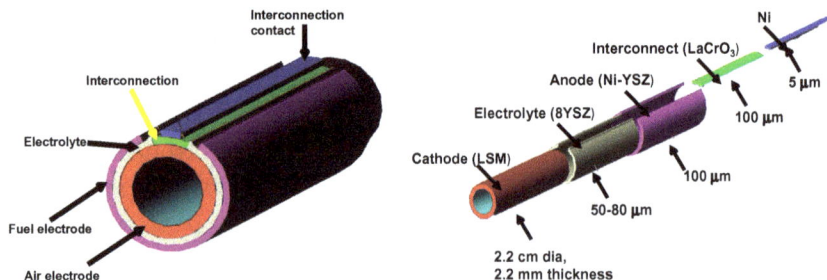

Figure 5.3. A typical Westinghouse tubular SOFC design drawing and the expanded view of its components along with their dimensions.

Metal-supported SOFCs are considered to offer an alternative to the conventional electrolyte and electrode-supported SOFCs (Nielsen *et al.*, 2018). They have many potential advantages such as good thermal conductivity and ductility of the metallic substrate, which may have improved thermal shock resistance, lowered internal temperature gradients, and quicker start-up (Nielsen *et al.*, 2018). In this design, low-cost alternative materials can be used as compared to traditional all ceramic-based devices. Additionally, they have different mechanical behavior, compared to ceramic components, and can withstand the mechanical stresses. However, the major issue holding back the growth of this technology is the poor densification of the electrolyte layer that usually requires high sintering temperatures (>1200°C).

In recent years, there has been a greater interest in the development of intermediate-temperature (IT) SOFCs, which operate in the range of 600°C–800°C. The greatest challenge has been the development of efficient cathode materials. Apart from LSCF, compounds with K_2NiF_4-type structure with inherent mixed ionic–electronic conductivity have attracted much attention because of their interesting structure, transport, and catalytic performance.

5.2.4. *Parameters Related to SOFC Performance*

The two important parameters that are measured for determining the SOFC are power density and area-specific resistance (ASR) and the formulae for calculating them are given below:

5.2.4.1. *Power density*

The power density of a cell is usually defined on the basis of the cell or electrode area, that is,

$$P_{\text{Cell}} = \frac{I(\text{current}) \times V(\text{voltage})}{A(\text{active area})}$$

The power density of a stack defined on the basis of stack volume is given by:

$$P_{\text{stack}} = \frac{I(\text{current}) \times V(\text{voltage})}{V(\text{stack volume})}$$

5.2.4.2. Area-specific resistance

The ASR of a fuel cell stack into which hydrogen (gas) and oxygen (air) are inputs, and electricity and exhaust gases are outputs, is defined as:

$$ASR = \frac{Emf - U}{i},$$

where Emf is the electromotive force with the inlet fuel and air, and U is the cell voltage at the current density, i, at the operating voltage.

5.2.5. *Methods of Synthesizing SOFC Materials*

The conventional way of synthesizing SOFC materials is by the solid-state reaction (SSR) with oxides. This method is not promising for many advanced applications because of the formation of large particles, agglomerates, chemical inhomogeneity, undesirable phases, abnormal grain growth, lower reproducibility, and imprecise stoichiometric control of cations. These materials can also be synthesized using wet chemical methods such as sol-gel, hydrothermal, chemical coprecipitation, solvothermal synthesis, and reverse micelle synthesis (Rahaman, 2006). The wet chemical synthesis routes yield highly reactive powders that can be sintered at a lower temperature (Rahaman, 2006). However, many of these methods require long reaction time, high external temperature, or special instrumentation. Among the many methods reported for the synthesis of SOFC materials, solution combustion (SC) method is very attractive and more details are presented in Section 5.3.

5.3. SC Method

The SC method has emerged as one of the most versatile methods for the synthesis of all types of oxide materials. The SC process involves the rapid heating of an aqueous redox mixture containing stoichiometric amounts of metal nitrates (oxidizer) and fuel in a preheated muffle furnace or a microwave oven (Patil *et al.*, 2008). The steps involved in the SC method are shown in the flowchart (Fig. 5.4). Many topical reviews and a book about this process have been published (Patil *et al.*, 2008; Patil *et al.*, 1997; Patil *et al.*,

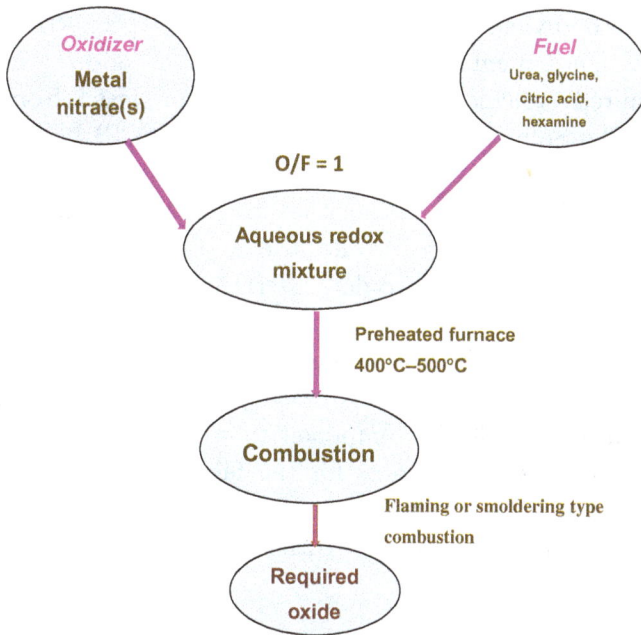

Figure 5.4. A flowchart depicting the steps involved in the SC synthesis of oxide materials.

2002; Aruna and Mukasyan, 2008; Varma *et al.*, 2016). This process is also referred to as autoignition, sol-gel combustion, citrate–nitrate, glycine–nitrate, flash combustion, and so on.

This method is very attractive because of the following salient features:

- It is a low-temperature-initiated, gas-producing, self-propagating combustion process yielding voluminous and fine ceramic oxides.
- It is a simple, fast, and economically attractive method as it employs readily available cheaper chemicals and does not require any sophisticated equipment.
- This process has the potential to synthesize any oxide with the desired composition, structure, morphology, and property for a specific application.
- It is very easy to dope or incorporate atomic-level metal ions in most of the oxide matrices.

The SC method yields oxide powders with unique properties required for SOFC components as any dopants can be added and oxides with required stoichiometry can be easily prepared from nitrate solution. The Solution Combustion Synthesis (SCS) has been used extensively for the synthesis of all kinds of SOFC materials. Only one review article specifically focuses on the application of the SCS materials for energy conversion and storage (Li *et al.*, 2015). However, it does not provide an up-to-date literature on the SCS materials pertaining only to SOFC (Li *et al.*, 2015). There is no comprehensive review or chapter on the SC-synthesized SOFC materials. To fill this lacuna, this chapter is focused toward providing an update on (1) the SC-synthesized SOFC component powders, (2) the application of the SC-synthesized powders for the fabrication of SOFC layers by ceramic processing techniques such as pressing and tapecasting, and (3) the application of SC-synthesized powders for developing coatings.

5.4. Solution Combustion-Synthesized SOFC Materials

In the 1990s, when the research on SOFC was in its infancy, the SC method was used for the preparation of all the SOFC component materials used as cathode, electrolyte, anode, and interconnect using hydrazide-based fuels (Aruna, 1998). The synthesis of the electrolyte and interconnect materials of SOFC using hydrazide fuels is discussed below.

In a typical synthesis, 8 mol% YSZ electrolyte was prepared using the combustion of an aqueous solution containing yttrium nitrate (stabilizer), zirconium oxynitrate, and carbohydrazide (CH_6N_4O) as the fuel in the required molar ratio. The added metal nitrate (stabilizer) decomposes into the respective metal oxide (MO), which reacts with metathetically formed ZrO_2 to give a solid solution, thereby stabilizing the high-temperature zirconia phase.

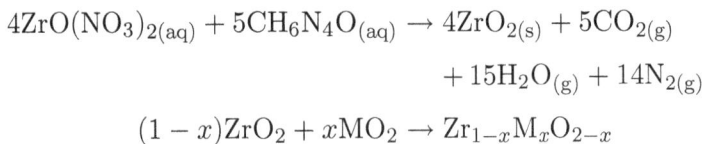

$$4ZrO(NO_3)_{2(aq)} + 5CH_6N_4O_{(aq)} \rightarrow 4ZrO_{2(s)} + 5CO_{2(g)}$$

$$+ 15H_2O_{(g)} + 14N_{2(g)}$$

$$(1-x)ZrO_2 + xMO_2 \rightarrow Zr_{1-x}M_xO_{2-x}$$

Table 5.5. Various SOFC components prepared by the SCS using hydrazide-based fuels and their properties in sintered pellet form.

SOFC components	Fuel used	Conductivity at 1173 K (Scm^{-1})	Thermal expansion coefficient at 1173 $(\times 10^{-6}\ K^{-1})$
Cathode			
$La_{0.84}Sr_{0.16}MnO_3$	Oxalyl dihydrazide	202	12.63
$La_{0.84}B_{a0.16}MnO_3$	Oxalyl dihydrazide	156	10.91
$Pr_{0.75}Sr_{0.25}MnO_3$	Tetraformal trisazine	114	10.2
$Nd_{0.75}Sr_{0.25}MnO_3$	Tetraformal trisazine	140	10.7
Anode			
30 vol% Ni/YSZ	Oxalyl dihydrazide	40	11.60
Electrolyte			
8 mol%YSZ	Carbohydrazide	0.11	10.70
Interconnect			
$La_{0.8}Ca_{0.2}Cr_{0.9}Co_{0.1}O_3$	Maleic hydrazide	23	11.65

Adapted from Aruna (1998).

Similarly, fine particles of lanthanum chromite powder were prepared from the combustion reaction of an aqueous redox mixture containing stoichiometric amounts of lanthanum nitrate, chromium nitrate, and hydrazide-based fuel (maleic hydrazide) at 350°C. The combustion reaction was of flaming type and yielded voluminous product and the formation of $LaCrO_3$ can be represented by the following equation:

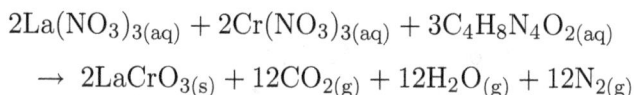

$$2La(NO_3)_{3(aq)} + 2Cr(NO_3)_{3(aq)} + 3C_4H_8N_4O_{2(aq)}$$
$$\rightarrow 2LaCrO_{3(s)} + 12CO_{2(g)} + 12H_2O_{(g)} + 12N_{2(g)}$$

Table 5.5 summarizes the different SOFC component materials prepared using the SC method using hydrazide-based fuels and their respective conductivity and thermal expansion coefficient values (Aruna, 1998).

All the SOFC components obtained by the SC method using hydrazide-based fuels were nanosized and exhibited a large surface area $(9\text{--}40\,\mathrm{m^2 g^{-1}})$ (Aruna, 1998). They were sinteractive and phase pure and satisfied the stringent requirements for SOFC application.

However, later hydrazide-based fuels were not used by researchers as they are not readily available and it is cumbersome to synthesize them. Instead, readily available fuels such as glycine, alanine, urea, citric acid, hexamethylenetetramine, sucrose, humic acid (HA), polyvinyl alcohol (PVA), and so on, were used. Several researchers have used the SC or autoignition method for the preparation of nanosized oxide powders of cathode, anode, and electrolyte using various fuels. Among the many fuels used, citric acid fuel has been used mainly because of the lower exothermicity of the combustion reaction with this fuel and thus, in turn, yields nanosized powders that are suitable for SOFC application (Magnone *et al.*, 2009). The various SOFC materials prepared using the SC method, fuel used, and the properties of the materials are summarized in Table 5.6.

5.5. SOFC Components and Single Cells from SC-Derived Powders

A large number of studies have been devoted to the synthesis of powders and processing of electrodes and electrolytes in SOFCs.

5.5.1. *SOFC Cathodes*

A large number of oxides have been prepared using the SC method and explored as cathode materials for SOFC. The oxides include $La_{0.8}Sr_{0.2}MnO_3$ (LSM), $La_{0.65}Sr_{0.3}MnO_3$ (LSM), $La_{0.6}Sr_{0.4}Co_{0.98}Ni_{0.02}O_3$, LNF, $Pr_{0.8}Sr_{0.2}Co_{0.5}Fe_{0.5}O_3$, LSCF, pristine $GdBaCo_2O_{5+\delta}$, $GdBaCo_{2/3}Fe_{2/3}Ni_{2/3}O_{5+\delta}$, $GdBaCo_{2/3}Fe_{2/3}Cu_{2/3}O_{5+\delta}$, $GdBaCoCuO_{5+\delta}$ (GBCO), Fe- and Cu-doped $SmBaCo_2O_{5+\delta}$(FC-SBCO)–$Ce_{0.9}Gd_{0.1}O_{1.95}$(CGO) composites, and $La_{0.5}Sr_{0.5}Co_{0.8}Fe_{0.2}O_3$. At the operating temperature of 1173 K, a relatively higher electrical conductivity and lower polarization resistance $(10.82\ \Omega\,\mathrm{cm^2})$ were exhibited by SC-synthesized LSM as against the solid-state method-synthesized LSM powder (Yang *et al.*, 2011). LSM was also prepared using urea and glycine as fuel and varying the fuel/oxidizer ratio. Excess urea fuel facilitated the formation of

Table 5.6. Various SOFC components prepared using the SC method reported in literature along with the fuel used and their properties.

Component	Composition	Fuel used	Property/Observations	Reference
Electrolyte	$La_{1-x}Sr_xGa_{1-y}Mg_y$ $O_{3-(x+y)/2}$	Glycine	Ionic conductivity higher than that of YSZ and TEC-11.3–12.1 × 10^{-6} K^{-1}	Stevenson *et al.* (1997)
	CeO_{2-x}, Gd_2O_3 GDC and Li- and Co-doped GDC	Citric acid	Conductivity 4.53 × 10^{-2}–1.26 × 10^{-1} Scm^{-1} in the temperature range of 600°C–800°C, which is suitable for IT-SOFCs.	Dutta *et al.* (2009c)
	$Ce_{0.8}Gd_{0.2}O_{2-\delta}$	Citric acid	Conductivity of ~1.74 × 10^{-2} Scm^{-1} at 600°C was obtained for material prepared from cerium nitrate under the fuel-deficient condition	
	10 mol% scandia-stabilized zirconia (10ScSZ)	Urea	10ScSz exhibited higher ionic conductivity compared to 8YSZ processed under the same conditions	Vijaya Lakshmi *et al.* (2011b)

(*Continued*)

Table 5.6.   (*Continued*)

Component	Composition	Fuel used	Property/ Observations	Reference
	Ytterbia co-doped ScSZ ($0.9ZrO_2$-$0.09Sc_2O_3$-$0.01Yb_2O_3$)	Glycine + PEG	Doped ScSZ exhibited lower activation energy (0.54 eV) compared to the undoped ScSZ (0.61 eV)	Vijaya Lakshmi et al. (2011a)
	$BaPr_{0.7}Gd_{0.3}O_{3-\delta}$	EDTA + acrylamide + N-NV-methyl-enbis acryl-amide + α-α'-azobutyronitrile	Crystallite size of 90 nm-high-temperature proton conductor solid electrolyte in IT-SOFC	Magrasó et al. (2004)
	$Ce_{1-x}Sm_xO_{2-x/2}$	Citric acid, alanine and glycine	Crystallite sizes: citric acid: 6–20 nm Alanine: 13–17 nm Glycine: 15–23 nm	Chung and Lee (2004)
	CaO-, Sm_2O_3-, Y_2O_3-, and Gd_2O_3-doped ceria	Urea	Homogeneous, and reactive powders	Chinarro et al. (2007)
	Li- and Co-doped GDC	Citric acid	Ionic conductivity of 4.53×10^{-2}–1.26×10^{-1} Scm^{-1} in the temperature range of 600°C–800°C	Accardo et al. (2018)

Table 5.6. (*Continued*)

Component	Composition	Fuel used	Property/ Observations	Reference
	Cu-Sm-doped CeO_2	PVA	The addition of 1 mol% Cu in $Ce_{0.8}Sm_{0.2}$ $O_{2-\delta}$ resulted in an improvement in electrical conductivity	Dong et al. (2011)
	$La_{0.9}Sr_{0.1}$ $Ga_{0.8}Mg_{0.115}$ $Co_{0.085}O_{2.85}$ (LSGMC)	Glycine	The sintered pellet exhibited conductivity values of 0.094 and 0.124 Scm^{-1} at 800°C and 850°C, respectively	Xue et al. (2010)
	TiO_2 and Al_2O_3 doped $(CeO_2)_{0.92}$ $(Y_2O_3)_{0.04}$ $(CaO)_{0.04}$	Urea	Titania addition resulted in a significant reduction in the grain-boundary resistance	Jurado (2001)
	Dy-doped $BaCeO_3$–$BaZrO_3$ $(BaCe_{0.5}Zr_{0.3}Dy_{0.2}$ $O_{3-\delta})$	Citric acid	Excellent electrical properties and chemical and thermal compatibility	(Lyagaeva et al. (2016)
	$BaZr_{0.8}Y_{0.2}$ $_{3-\delta}$	Citric acid	Higher proton conductivity for SCS-synthesized powder compared to solid state-synthesized powder	Peng et al. (2014)

(*Continued*)

Table 5.6. (Continued)

Component	Composition	Fuel used	Property/ Observations	Reference
Cathode	$La_{1-x}Sr_x$ MnO_3 ($x = 0$, 0.1, 0.16, 0.2, and 0.3)	Oxalyl dihydrazide	$La_{0.84}Sr_{0.16}$ MnO_3-Conductivity- 202 Scm^{-1} and TEC -12.638×10^{-6} K^{-1} at 900°C	Aruna et al. (1997)
	$LaNi_{0.6}$ $Fe_{0.4}O_3$ (LNF)	Citric acid and urea	TEC of LNF samples at 800°C was 11.8×10^{-6} K^{-1} and a conductivity of 344 Scm^{-1} at 800°C observed for LNF	Basu et al. (2004)
	$Pr_{0.8}Sr_{0.2}$ $CO_{1-x}Fe_xO_3/$ $Pr_{0.8}Sr_{0.2}$ $Co_{0.5}Fe_{0.5}$ $O_{3-\delta}$	Citric acid	Resistance of symmetrical cells evaluated	Magnone et al. (2010)
	$Pr_{0.8}Sr_{0.2}$ $Co_{0.5}Fe_{0.8}$ $O_{3-\delta}$	Citric acid	ASR value of 2.36 Ωcm^2 at 973 K	Magnone et al. (2006)
	$LaMnO_3$-$LaCoO_3$	Tetraformal trisazine	Composition with very low Co content ($x < 0.2$) suitable for SOFC application	Aruna et al. (2000)

Table 5.6.   (*Continued*)

Component	Composition	Fuel used	Property/ Observations	Reference
	$La_{1-a}Sr_a$ Co_{1-b} Fe_bO_{3-x} ($a = 0.3–0.5$; $b = 0–0.2$), LSCF	Citric acid	Best composition with $a = 0.4$ and minimum iron doping exhibited ASR of $0.13\ \Omega \cdot cm^2$ at $700°C$	Deganello *et al.* (2005)
	$GdBaCo_{2/3}$ $Fe_{2/3}Cu_{2/3}$ $O_{5+\delta}$	Citric acid	Best cathode material for IT-SOFC was 0.666 equivalent-molar composition of Fe- and Cu-co-doped GBCO	Jo *et al.* (2009)
	$La_{0.6}Sr_{0.4}$ $Co_{0.98}Ni_{0.02}O_3$	Glycine	Highest electrical conductivity of 96 Scm^{-1} at $800°C$ in air	Dutta *et al.* (2009a)
	$(La_{0.75}Sr_{0.25})$ $Cr_{0.5}Mn_{0.5}$ $O_{3-\delta}$	Ethylene glycol	Symmetrical SOFC was demonstrated	Bastidas *et al.* (2006)
	Sr-doped La_2CuO_4	Citric acid	Promising cathode material for GDC-based IT-SOFC	Caronna *et al.* (2010)
	$La_{0.6}Sr_{0.4}$ $Co_{0.2}Fe_{0.8}O_3$ (LSCF)	Cellulose	Particle size: 15–20 nm	Cai *et al.* (2010)

(*Continued*)

Table 5.6. (*Continued*)

Component	Composition	Fuel used	Property/ Observations	Reference
	Fe, Cu-SmBaCo$_2$ O$_{5+\delta}$–Ce$_{0.9}$ Gd$_{0.1}$O$_{1.95}$	Citric acid	Composite cathode	Lee *et al.* (2011)
	Nd$_{2-x}$Sr$_x$ NiO$_{4+\delta}$	Strontium acetate, neodymium acetate	Cathode for IT-SOFC	Khandale *et al.* (2013)
Anode	Ni$_{1-x}$ Co$_x$/YSZ	Urea	Porous green compacts by uniaxial or isostatic pressing Symmetrical cells cermet/ electrolyte/cermet, lower polarization resistance	Ringuedé *et al.* (2002)
	NiO/Ce$_{0.8}$ Sm$_{0.2}$O$_{1.9}$ (SDC)	Polyacrylamide hydrogel and HA	The single-cell SOFCs exhibited a maximum power density of 740 mWcm^{-2} at 650°C	Dong *et al.* (2010)
	Ni-YSZ, Ni,Co–YSZ, Ni,Fe–YSZ, and Ni, Cu–YSZ	Urea	Nanometric and submicronic particles, with high-specific surface area	Ringuede *et al.* (2001)

Table 5.6. (*Continued*)

Component	Composition	Fuel used	Property/ Observations	Reference
	$NiO-Ce_{0.9}Gd_{0.1}O_{1.95}$	Citric acid	NiO–10GDC anode-coated semi-SOFC cell exhibited an electrical conductivity of the order of 103 Scm^{-1} at 600°C with respect to temperature	Akbari-Fakhrabadi *et al.* (2012)
	$Ni-SrZr_{0.95}Y_{0.05}O_{2.975}$ $Ni-CaZr_{0.95}Y_{0.05}O_{2.975}$ and $Ni-SrCe_{0.475}Zr_{0.475}Y_{0.05}O_{2.975}$	Urea	$Ni-SrZr_{0.95}Y_{0.05}O_{2.975}$, no interfacial reaction product was observed	Mather *et al.* (2004)
	$Ce_{0.6}Mn_{0.3}Fe_{0.1}O_2$ (CMF)	Glycine	Screen printed as anode	Zhu *et al.* (2013)
Interconnect	$LaCrO_3$ and $La_{0.8}Ca_{0.2}Cr_{0.7}Co_{0.3}O_4$	Urea	Microwave used for ignition Surface area = 12.9 m^{2}g^{-1} and 22.6 m^{2}g^{-1}	Kiminami and Morelli, (2004)
	Sr-, Co-doped $LaCrO_3$	Urea	The synthesized powder was suitable for tapecasting	Setz *et al.* (2010)
	$La_{0.8}Sr_{0.2}Cr_{0.92}Co_{0.08}O_3$	Urea	Zetapotential studies	Setz *et al.* (2009)

pure LSM phase, whereas $SrCO_3$ and $MnCO_3$ phases were found as impurities along with LSM when glycine was used as the fuel (da Conceição *et al.*, 2009). $La_{1-x}Sr_xMnO_3$ was synthesized using a unique combination of oxidant and fuel and nitrate–acetate in a stoichiometric ratio. Manganese acetate acted as fuel. However, a calcination step at 600°C was adopted to eliminate carbon residues. This process is useful to produce nanosized LSM powder in large scale because of a gradual and smooth combustion and hence the process is devoid of any fire hazards or explosions (Saha *et al.*, 2006).

The properties of LNF powder synthesized using urea and citric acid fuels were compared with LSM to explore LNF as a prospective cathode material for SOFC (Basu *et al.*, 2004). LNF powders synthesized using urea and sintered at 1400°C exhibited a conductivity of 344 Scm^{-1} at 800°C, which is higher than the state-of-the-art LSM that exhibits a conductivity of only 228 Scm^{-1} at 800°C. Also, LNF sample retained a porosity of around 20% even after firing at 1400°C and thus appears to be a promising SOFC cathode component (Basu *et al.*, 2004).

Nanocrystalline 2 mol% Ni-doped lanthanum strontium cobaltite ($La_{0.6}Sr_{0.4}Co_{0.98}Ni_{0.02}O_3$) synthesized by the SC method using slightly higher glycine fuel ratio (0.45) displayed the highest electrical conductivity of 96 Scm^{-1} at 800°C in air (Dutta *et al.*, 2009a). This material was sintered at 1000°C and used as the cathode in the symmetric cell mode along with Gd_2O_3-doped CeO_2 as the electrolyte. The fabricated cell exhibited a power density of ~ 2.4 Wcm^{-2} at 800°C (Dutta *et al.*, 2009a).

Porous cathodes were fabricated from nanoparticles of $Sm_{0.5}Sr_{0.5}CoO_3$ (SSC) prepared by the SC method using a new fuel such as HA (Khandale *et al.*, 2013). The single-cell SOFCs made from these nanoparticles exhibited a maximum power density of 740 mWcm^{-2} at 650°C (Dong *et al.*, 2010).

$Pr_{0.8}Sr_{0.2}Co_{0.5}Fe_{0.5}O_3$ synthesized using citric acid as fuel was also explored as the cathode material (Magnone *et al.*, 2010). In another study, orthorhombic-type $Pr_{0.8}Sr_{0.2}Co_{0.2}Fe_{0.8}O_{3-\delta}$ ($x = 0.8$) phase synthesized by the SC method exhibited an ASR of 2.36 Ωcm^2 at 973 K (Magnone *et al.*, 2009). Combustion-synthesized $Pr_{0.8}Sr_{0.2}Co_{1-x}Fe_xO_{3-\delta}$ powder was applied on both sides of $Ce_{0.8}Sm_{0.2}O_{2-}$pellets to form a symmetric cell and the electrochemical properties of the electrode/electrolyte interfaces as a function of the temperature, Fe content (x), and oxygen partial pressure were studied using electrochemical impedance spectroscopy (EIS) (Magnone *et al.*, 2006).

LSCF perovskite oxide prepared by the SC method using cellulose as the fuel showed higher purity and better cathode performance in SOFC (Cai *et al.*, 2010). This was attributed to the suppression of $SrCO_3$ impurity in the LSCF product with cellulose as the fuel. LSCF was also prepared using glycine as fuel and was applied as a coating on YSZ and GDC pellets and sintered at 1050°C for 2 h. It showed a higher polarization resistance of 42 Ωcm^2 at 700°C because of the formation of an insulating interfacial layer during the high-temperature fabrication step (Shri Prakash *et al.*, 2016).

Chanquía *et al.* (2014) synthesized pure-phase $La_{0.4}Sr_{0.6}$ $Co_{0.8}Fe_{0.2}O_{3-\delta}$ (LSCF) nanocrystallites ($\sim$45 nm and surface area 10 m^{2}g^{-1}) with a sponge-like structure consisting of meso- and macropores using the SCS method and employed glycine as fuel and ammonium nitrate as combustion trigger. Symmetrical cells were fabricated by the spin coating method using LSCF and $La_{0.8}Sr_{0.2}Ga_{0.8}Mg_{0.2}O_{3-\delta}$. The ASR of the nanostructured sample (45 nm) decreased by two orders of magnitude with respect to the sub-microstructured sample (685 nm), reaching values as low as 0.8 Ω cm^2 at 450°C. This was attributed to the optimized nanoscale cathode morphology, which in turn resulted in shortening of the oxygen diffusion paths and reduced polarization resistance associated with the surface exchange and O-ion bulk diffusion process (Chanquía *et al.*, 2014).

Nanocrystalline LSCF powder (0.9 μm average agglomerated particle size, 74-nm particle size, and 27-nm crystallite size) required for screen printing was synthesized through the SC method utilizing higher glycine fuel (fuel/oxidizer = 2) followed by calcination at 900°C. A symmetrical cell with the following configuration LSCF/$Ce_{0.9}Gd_{0.1}O_{1.95}$ (GDC)/LSCF was studied to understand the electrode/electrolyte interface. To evaluate the potential application of LSCF in SOFCs, the temperature-programmed reduction and oxidation were performed on the LSCF. A strong reduction and oxidation behavior were observed, respectively, around $860°C$ and $388°C$. The symmetrical cell displayed a minimum charge transfer resistance of $6.3 \ \Omega cm^2$ at $550 \ °C$ (Jamale *et al.*, 2015).

There has been an increased trend in employing composite cathodes to improve the performance of SOFCs. $Ba_{0.5}Sr_{0.5}Co_{0.8}Fe_{0.2}O_{3-\delta}$-$La_{0.9}Sr_{0.1}Ga_{0.8}Mg_{0.2}O_{3-\delta}$ composite cathodes were prepared by the SCS method using glycine as fuel. The addition of $La_{0.9}Sr_{0.1}Ga_{0.8}Mg_{0.2}O_{3-\delta}$ facilitated the suppression of grain growth and grain sizes decreased with an increasing wt% of $La_{0.9}Sr_{0.1}Ga_{0.8}Mg_{0.2}O_{3-\delta}$ phase in the composites. There was excellent chemical compatibility between $La_{0.9}Sr_{0.1}Ga_{0.8}Mg_{0.2}O_{3-\delta}$ and $Ba_{0.5}Sr_{0.5}Co_{0.8}Fe_{0.2}O_{3-\delta}$ when the wt% of $La_{0.9}Sr_{0.1}Ga_{0.8}Mg_{0.2}O_{3-\delta}$ in the composite was not more than 40% (Liu *et al.*, 2008).

Symmetric SOFC was fabricated from SC-synthesized nanostructured $La_{1-a}Sr_aCo_{1-b}Fe_bO_{3-\delta}$ ($a = 0.3$–$0.5; b = 0$–0.2) and $Ce_{0.8}Sm_{0.2}O_2$ powders. An ASR of $0.13 \ \Omega cm^2$ at $700°C$ was observed for LSCF composition with $a = 0.4$ and the minimum iron doping (Deganello *et al.*, 2005). A nanocomposite cathode comprising of Fe-rich LSCF and Sm-Gd co-doped ceria-rich $Ce_{0.8}Sm_{0.1}Gd_{0.1}O_{1.90}$ (CSGO) in the ratio of 1:1 exhibited a total electrical conductivity of $0.043 \ Scm^{-1}$ at $700°C$, which is higher than that of the LSCF composite containing singly doped compositions (Kumar *et al.*, 2018). Similarly, nanopowders of $Sm_{0.5}Sr_{0.5}CoO_{3-\delta}$ and $La_{0.6}Sr_{0.4}CoO_{3-\delta}$ (LSC) compositions synthesized using glycine fuel were investigated as cathode materials for intermediate temperature SOFCs (Bansal and Zhong, 2006).

Transition-metal-doped double-perovskite oxides such as $GdBaCo_{2/3}Fe_{2/3}Ni_{2/3}O_{5+\delta}$ (FN-GBCO), $GdBaCo_{2/3}Fe_{2/3}Cu_{2/3}O_{5+\delta}$ (FC-GBCO), $GdBaCoCuO_{5+\delta}$ (C-GBCO) and pristine $GdBaCo_2O_{5+\delta}$ (GBCO) were synthesized via citrate SC method. ES single-cell (FC-GBCO/CGO/Ni-CGO) fabricated from the synthesized powders appeared to be a potential cathode material for IT-SOFCs as FC-GBCO cathode exhibited reduced thermal expansion and high electrochemical performance in the IT regime of 700°C with the CGO electrolyte (Jo *et al.*, 2009).

A composite cathode consisting of Fe- and Cu-doped $SmBaCo_2O_{5+\delta}$ (FC-SBCO)–$Ce_{0.9}Gd_{0.1}O_{1.95}$ (CGO)(0–40 wt%) synthesized using citrate combustion method was explored as a new cathode material for IT-SOFC (Lee *et al.*, 2011). The optimized cell consisting of composite cathode of 30 wt% CGO–70 wt% FC-SBCO (CS30-70) coated on a $Ce_{0.9}Gd_{0.1}O_{1.95}$ electrolyte exhibited the lowest ASR of 0.049 Ωcm^2 at 700°C. A maximum power density of 535 $mWcm^{-2}$ at 700°C was exhibited by the ES (300 μm thick) single cell with the configuration of CS30-70/CGO/Ni-CGO (Lee *et al.*, 2011).

Well-dispersed $Pr_{0.35}Nd_{0.35}Sr_{0.3}MnO_{3-\delta}$(PNSM)/YSZ composite cathode powders suitable for tubular SOFCs were prepared by microwave-induced monomer gelation and gel combustion synthesis (Dong *et al.*, 2008). The cathode powder was slurry sprayed on an extruded NiO-YSZ tube containing the dense YSZ film and sintered at 1100°C for 2 h and the obtained cell is shown in Fig. 5.5 (Dong *et al.*, 2008).

The electrochemical performance of YSZ-based planar SOFC with CGO interlayer was studied by Dutta *et al.* (2009c) and the

Figure 5.5. Photograph of the as-prepared cell. Adapted from Dong *et al.* (2008).

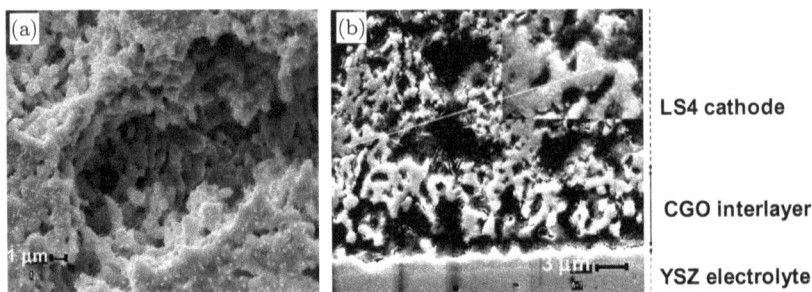

LS4 cathode

CGO interlayer

YSZ electrolyte

Figure 5.6. SEM images of (a) the top layer of $La_{0.5}Sr_{0.5}Co_{0.8}Fe_{0.2}$(LS4) cathode and (b) corresponding fractured anode-supported single cell with CGO as interlayer after the cell test. Adapted from Dutta *et al.* (2009b).

performance varied with the composition of the developed cathodes. The highest current density of ~ 1.72 Acm^{-2} and power density of 1.2 Wcm^{-2} at 0.7 V were obtained at $800°C$ with SC-derived $La_{0.5}Sr_{0.5}Co_{0.8}Fe_{0.2}O_3$ (LS4) powder synthesized using alanine as the cathode in anode-supported cell configuration with CGO as the interlayer (Dutta *et al.*, 2009b). Also, the lowest value of total ASR was observed (~ 0.211 Ωcm^2) for the cells with LS4 cathode only. The improved electrochemical performance of the developed anode-supported single cells was attributed to the small nanocrystalline growth inside the core grains (Fig. 5.6) (Dutta *et al.*, 2009b).

5.5.2. *SOFC Anodes*

In this section, the details of the anode materials synthesized using the SC method are presented. Nanocrystalline $NiO/Ce_{0.9}Gd_{0.1}O_{1.95}$ (10GDC) composite powder synthesized by the SC method using citric acid as fuel was formed as a cell. The obtained semi-SOFC cell exhibited an electrical conductivity of the order of 103 Scm^{-1} at $600°C$ (Akbari-Fakhrabadi *et al.*, 2012). In recent years, there is a great interest in developing new anode materials suitable for low- and intermediate-temperature SOFCs so as to improve the fuel cell lifetime and performance. Toward achieving this goal, new nanocrystalline anode powders with Mo_y-$M_xCe_{1-x}O_{2-\delta}$ (M: Ni, Co, and Cu) compositions were synthesized using glycine–nitrate SCS

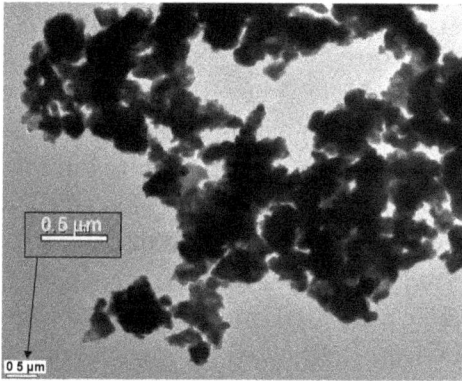

Figure 5.7. TEM image of Mo_y-$M_xCe_{1-x}O_{2-\delta}$ (Ce, ($Ni_{0.45}$, $Co_{0.35}$, $Cu_{0.20}$) powders heated at $900°C$. Adapted from Yildiz *et al.* (2015).

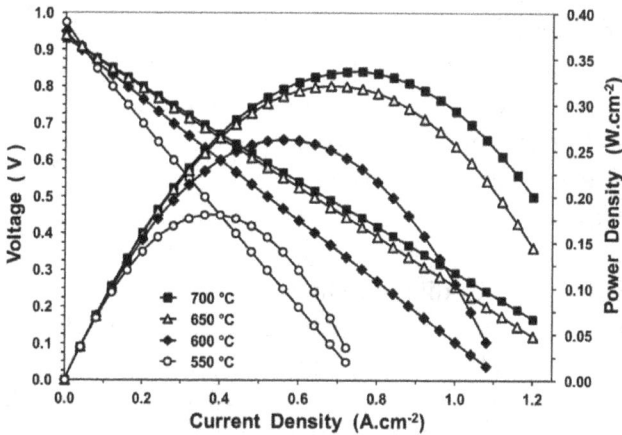

Figure 5.8. Single-cell performance analysis of Mo_y-$M_xCe_{1-x}O_{2-\delta}$ (Ce, ($Ni_{0.45}$, $Co_{0.35}$, $Cu_{0.20}$) at different working temperatures. Adapted from Yildiz *et al.* (2015).

process (Fig. 5.7) (Yildiz *et al.*, 2015). ES single cells with the configuration of LSM/GDC/ceria-based anode produced from the aforementioned nanopowders showed a power density of 0.35 Wcm^{-2} at $700°C$ under 25 mL min^{-1} dry CH_4 (Fig. 5.8) (Yildiz *et al.*, 2015). The developed anode can be a potential candidate for hydrocarbon-based SOFCs.

Three-layered symmetrical cells consisting of SC-synthesized $(Ni_{1-x}Co_xO_y)$/YSZ cermet/YSZ/$(Ni_{1-x}Co_xO_y)$/YSZ were fabricated by co-firing at 1450°C, followed by reduction to obtain porous $Ni_{1-x}Co_x$/YSZ cermet layers with good adhesion to the electrolyte. The results also showed that partial substitution of Ni with Co lowered the polarization resistance and this trend was reverted for high fractions of Co (Ringuedé *et al.*, 2002).

Metal-supported (MS) SOFCs have the potential to decrease material cost when compared to electrode-supported designs and usually MS-SOFCs are fabricated by the plasma spray technique. This technique requires flowable powders to form the electrolyte and other electrodes. Plasma spray grade flowable NiO-YSZ anode powders were prepared by the SCS using hexamine as fuel in a single step without involving the conventional agglomeration process of spray drying (Shri Prakash *et al.*, 2016). The Ni/YSZ anode consisting of nano YSZ particles, nano Ni particles, and nanopores was achieved on reducing the plasma-sprayed coating (Fig. 5.9). The Ni-YSZ coating exhibited 27% porosity adequate for anode functional layers and electronic conductivity of the coatings was $\sim$600 Scm^{-1} at 800°C (Shri Prakash *et al.*, 2016).

Using glycine as fuel, $Ce_{0.6}Mn_{0.3}Fe_{0.1}O_2$ (CMF) was prepared by the SC method using different fuel/oxidizer ratios (CMF1, CMF2, and CMF3 at $\Phi = 0.5$, $\Phi = 1$, and $\Phi = 1.8$, respectively) and was explored as the anode material (Zhu *et al.*, 2013). The as-synthesized powders were used for preparing anodes by sintering the screen-printed powders on the electrolyte membrane to fabricate single cells with $CMF/La_{0.8}Sr_{0.2}Ga_{0.8}Mg_{0.15}Co_{0.05}O_3$/SSC configuration using humidified hydrogen gas as a fuel and O_2 as the oxidizer. In the same configuration, cells were fabricated using powders synthesized using SSR method and the maximum power densities obtained were 1.23 and 1.09 Wcm^{-2} at 1000°C, respectively, for the cells fabricated from SCS- and SSR-derived powders, respectively, under the same evaluation conditions (Fig. 5.10) (Zhu *et al.*, 2013). This clearly shows the improved performance of SOFC fabricated from the SCS-derived powders. This may be attributed to the large surface area and nanosized grains of the SCS-derived powders.

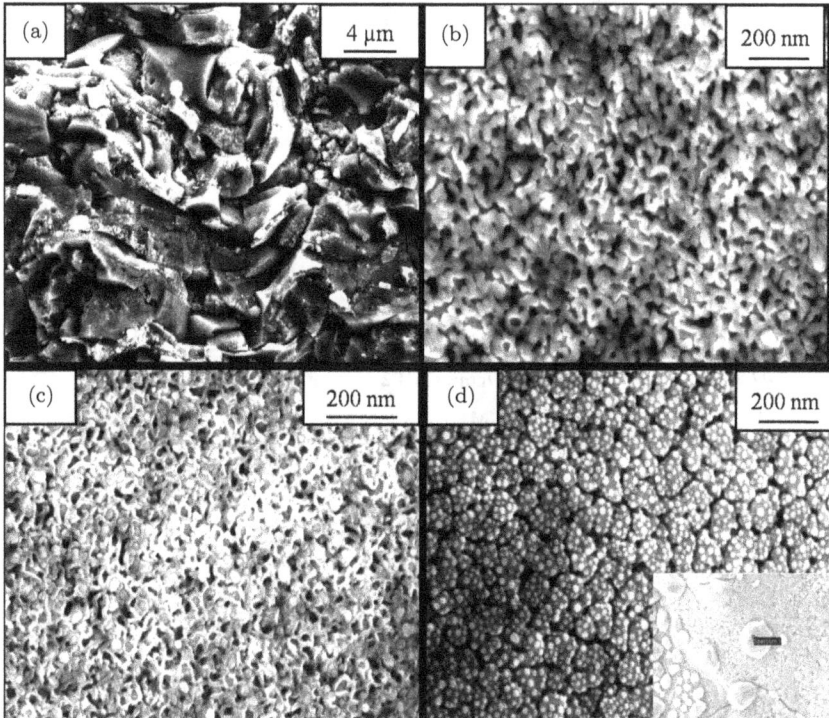

Figure 5.9. (a) Cross-sectional micrograph of Ni/YSZ coating deposited at
36 kW. High magnification micrographs of the coating deposited at (b) 24, (c) 30,
and (d) 36 kW. Adapted from Shri Prakash *et al.* (2016).

5.5.3. *SOFC Electrolyte*

There has been a renewed interest in the preparation of different electrolytes suitable for intermediate-temperature SOFCs. The important electrolyte materials studied are YSZ, scandia-stabilized zirconia
(ScSZ), GDC, lanthanum strontium gallium manganite (LSGM), and
samaria-doped ceria (SDC) and the details are presented below.
YSZ powders [$(xY_2O_3 - (1 - x)ZrO_2)$, $x = 5$, 8, and 10 mol%]
were synthesized using coprecipitation, sol-gel combustion, and sol-
gel methods. Sol-gel combustion-synthesized powder exhibited the
highest ionic conductivity (Zarkov *et al.*, 2015).

By varying the glycine fuel amounts in the aqueous redox
mixture, alkaline earth-doped lanthanum gallate powders were

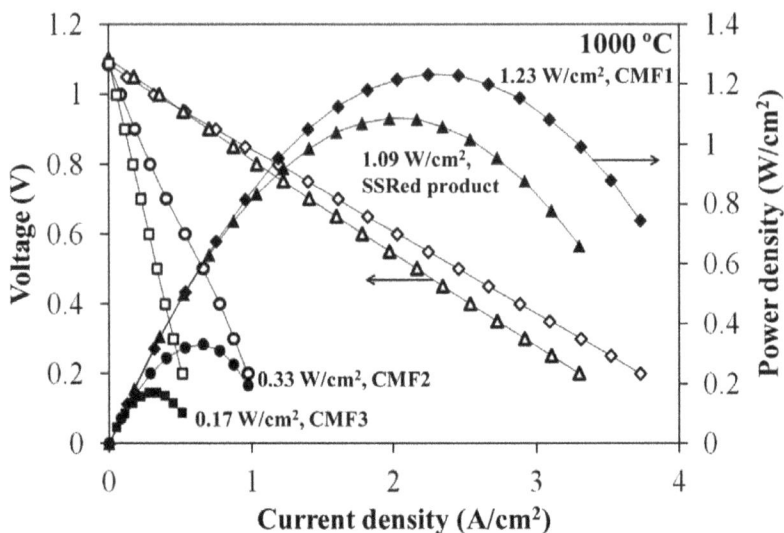

Figure 5.10. Power density curves of cells with CMF as the anode at $1000°$C with $La_{0.8}Sr_{0.2}Ga_{0.8}Mg_{0.15}Co_{0.05}O_3$ as the electrolyte and SSC as cathode. Adapted from Zhu *et al.* (2013).

synthesized. Powder compacts synthesized from powders synthesized under fuel-rich conditions on sintering achieved densities greater than 97% of theoretical density. Interestingly, by doping A-site of the perovskite structure with Sr or Ba, and Mg on the B-site, the oxygen ion conductivity of lanthanum gallate could be enhanced, which was higher than that of YSZ. However, by doping with Ca and Mg, the conductivity of lanthanum gallate reduced to lower than that of YSZ. Also, the thermal expansion coefficients of the doped gallates were higher than that of YSZ (Stevenson *et al.*, 1997).

ScSZ is another promising electrolyte that possesses higher ionic conductivity than YSZ. The ionic conductivity of SC-synthesized 10 mol% ScSZ prepared from urea as fuel was found to be higher than 8YSZ processed under similar conditions (Vijaya Lakshmi *et al.*, 2011b). Ytterbia co-doped ScSZ was prepared by the SCS using glycine as fuel. Interestingly, it was found that sintered 10ScSZ contained an additional phase along with the cubic phase, whereas a single cubic phase was found in the ytterbia co-doped $0.9ZrO_2$-$0.09Sc_2O_3$-$0.01Yb_2O_3$ sample. This was reflected in the higher ionic conductivity and lower activation energy compared to 10ScSZ (Vijaya Lakshmi *et al.*, 2011a).

GDC ($Gd_{0.15}Ce_{0.85}O_2$-(GDC)) is a promising electrolyte material for intermediate-temperature SOFC and this material has been prepared considerably by the SC method. The equation for the formation of single-phase nanocrystalline $Ce_{0.9}Gd_{0.1}O_{1.95}$ (GDC10) powders by combustion technique using glycine as fuel may be written as follows (Singh *et al.*, 2011):

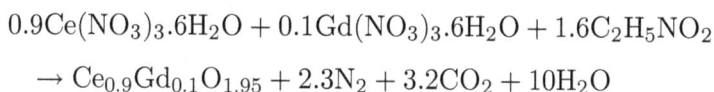

$$0.9Ce(NO_3)_3.6H_2O + 0.1Gd(NO_3)_3.6H_2O + 1.6C_2H_5NO_2$$
$$\rightarrow Ce_{0.9}Gd_{0.1}O_{1.95} + 2.3N_2 + 3.2CO_2 + 10H_2O$$

The nanocrystalline GDC showed low-temperature sinterability and thus could be beneficial in cofiring of the electrolytes and the electrodes (Jadhav *et al.*, 2009). GDC electrolyte with the composition $Ce_{0.85}Gd_{0.15}O_{2-\delta}$ synthesized using citric acid as fuel could be sintered to $> 95\%$ of the theoretical density at $1300°C$ for 10 h and exhibited an enhanced conductivity value (Singh *et al.*, 2011). Thus, SCS-synthesized GDC satisfies the stringent requirements of an IT-SOFCs solid electrolyte. It has been observed that the presence of small amounts of titania in the SC-synthesized CGO is beneficial as it produces a significant reduction of the grain-boundary resistance (Jurado, 2001).

The versatility of the SC method for the preparation of GDC electrolyte powders of varying characteristics suitable for various fabrication techniques such as wet powder spraying or tapecasting or plasma spraying has been demonstrated by varying the fuels (Shri Prakash *et al.*, 2014). GDC powder prepared by SCS using oxalyl dihydrazide fuel could be sintered to near theoretical density and also exhibited high conductivity (3×10^{-4} Scm^{-1} at $400°C$). These powders exhibited the necessary characteristics suitable for wet powder spraying (Fig. 5.11). Interestingly, the powders obtained by the SCS using a reduced amount of glycine and additional ammonium acetate-are of micron-sized, plasma-sprayable grade GDC. These

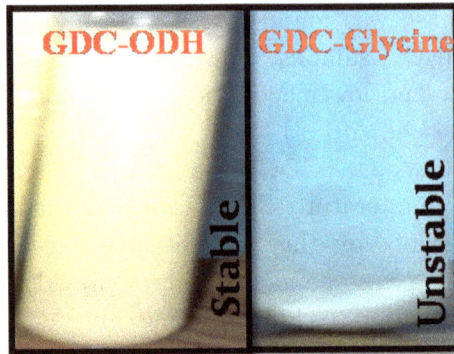

Figure 5.11. Photograph showing stable suspension of GDC-ODH powder as against unstable GDC-glycine suspension. Adapted from Shri Prakash *et al.* (2014).

powders were successfully plasma sprayed using optimized spray parameters and dense GDC electrolyte layer was obtained (Fig. 5.12) (Shri Prakash *et al.*, 2014).

GDC ($Ce_{0.9}Gd_{0.1}O_{1.95}$, GDC) electrolyte powder was synthesized using citrate-based SCS (Akbari-Fakhrabadi *et al.*, 2012). Interestingly, the as-synthesized powders possessed appropriate particle size and distribution suitable for aqueous-based tape casting method [80]. The GDC green tape possessed a smooth surface, flexibility, thickness in the range of 0.35–0.4 mm, and 45% relative green density (Fig. 5.13). It was demonstrated that flash sintering is a viable option for densifying the ceramics by surpassing sintering at high-temperatures and retarding the grain growth (Akbari-Fakhrabadi *et al.*, 2012).

Dong *et al.* (2011) synthesized nanosized $Ce_{0.8}Gd_{0.2}O_{2-\delta}$ and $Ce_{0.79}Gd_{0.2}Cu_{0.01}O_{2-\delta}$ electrolyte powders using PVA as fuel. By doping Cu in $Ce_{0.8}Gd_{0.2}O_{2-\delta}$, a significant reduction in densification temperature ($\sim$400°C) was obtained. Also, $Ce_{0.79}Gd_{0.2}Cu_{0.01}O_{2-\delta}$ exhibited higher conductivity (0.026 Scm^{-1}) compared to $Ce_{0.8}Gd_{0.2}O_{2-\delta}$ (0.0065 Scm^{-1}) at 600°C. $Ce_{0.79}Gd_{0.2}Cu_{0.01}O_{2-\delta}$ also exhibited good mechanical performance with three-point flexural strength value of 148.15 $\pm$ 2.42 MPa (Moreno *et al.*, 2014). $Ce_{0.9}Gd_{0.1}O_{2-\delta}$ (10CGO) prepared using different fuels such as urea,

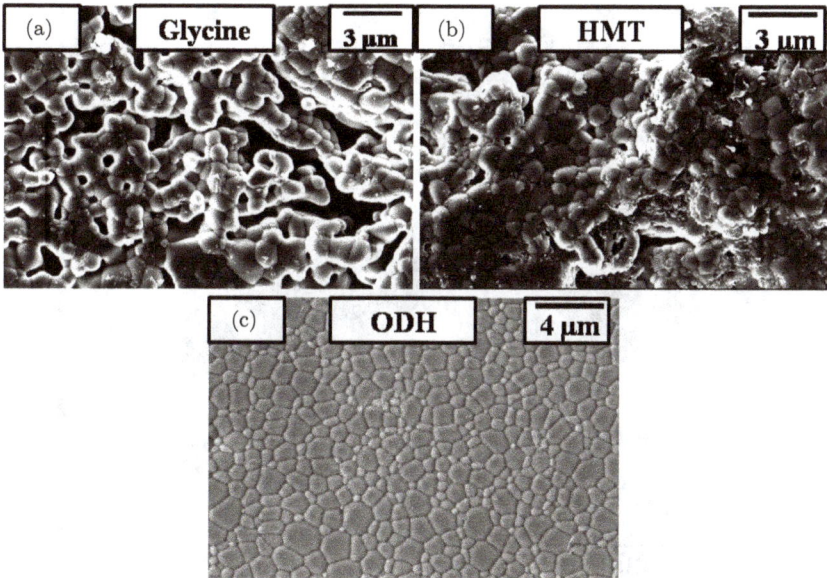

Figure 5.12. FESEM images of GDC pellets (sintered at 1300°C/5 h) prepared from powders synthesized using different fuels. Adapted from Shri Prakash *et al.* (2014).

Figure 5.13. Photograph of a green GDC tape fabricated from the SCS powder. Adapted from Akbari-Fakhrabadi *et al.* (2012).

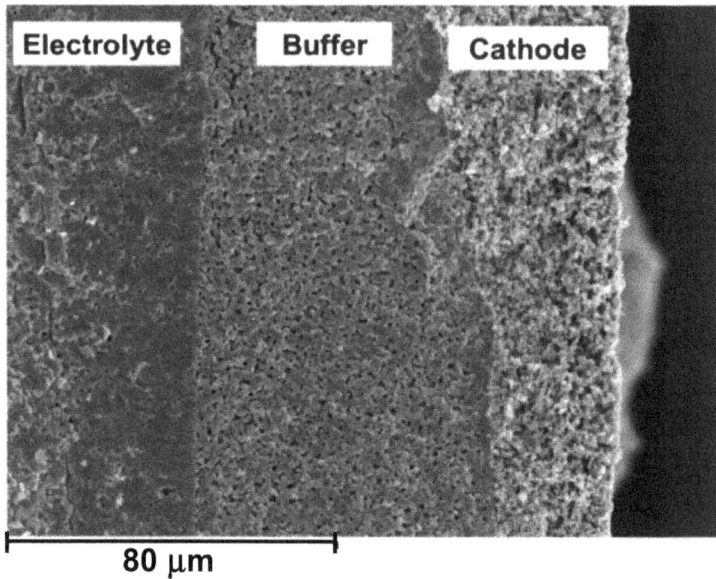

Figure 5.14. Symmetrical cell prepared by green tape forming the electrolyte (YSZ tape casting) and the buffer layer (10GDC, combustion synthesis), followed by the cathode (LSCF). Adapted from Moreno *et al.* (2014).

sucrose, and citric acid was used for processing symmetrical cells that allow the electrochemical characterization of CGO interface. (Moreno *et al.*, 2014). This work demonstrated the feasibility of using the SC-synthesized 10CGO powder synthesized using urea as fuel as a buffer layer to produce a functional symmetrical cell (Fig. 5.14) (Moreno *et al.*, 2014).

To improve the microstructural and electrochemical properties of the SCS-derived GDC, it was co-doped with 0.5–2 mol% of lithium and cobalt oxides (Accardo *et al.*, 2018). 2LiGDC and 0.5CoGDC exhibited a conductivity of 1.26×10^{-1} Scm^{-1} and 8.72×10^{-2} Scm^{-1} at 800°C after sintering at 1000°C and 1100°C, respectively (Accardo *et al.*, 2018). Nanopowders of $Ce_{1-(x+y)}Gd_xCa_yO_{2-(0.5x+y)}$ were successfully synthesized using glycine as fuel (Ainirad *et al.*, 2011). Ceria doped with yttria and calcia with the composition $Ce_{0.8+x}Y_{0.2+2x}Ca_xO_{1.9}$ was synthesized by the low-temperature combustion synthesis method using citric acid as fuel and ammonium hydroxide and ammonium nitrate as combustion aids (Xu *et al.*, 2008). By co-doping ceria with Y^{3+} and Ca^{2+}, the conductivity of ceria was enhanced (Xu *et al.*, 2008).

Recently, SDC is drawing attention as an alternative electrolyte to YSZ because of its low-temperature ionic conductivity and negligible chemical reactivity with the cathode materials. $Ce_{0.8}Sm_{0.2}O_{2-x}$ powders were synthesized by the SC synthesis using three different fuel mixtures. The oxide prepared with sucrose displayed higher specific surface area and a total pore volume compared to the samples synthesized with citric acid and cellulose-citric acid resulting in marked reducibility in H_2 at low temperature (Aliotta *et al.*, 2016). SDC powders were also synthesized using microwave-induced heating combustion process using citric acid, alanine, and glycine as the fuels (Fig. 5.15). Citric acid synthesized SDC powder exhibited lower crystallite size (Chung and Lee, 2004).

By using PVA as fuel in the SC method, SDC ($Ce_{0.8}Sm_{0.2}O_{2-\delta}$) powders were synthesized at a lower temperature and SSC was synthesized using glycine–nitrate method (Jiang *et al.*, 2007). A thin ($\sim$10 μm) SDC electrolyte film was prepared on a porous NiO-SDC anode substrate by the dry pressing method and a single cell with the following configuration NiO-SDC/SDC/SSC-SDC was fabricated and tested with humidified hydrogen and air as fuel and oxidant, respectively. The cell exhibited a power density of 936 mWcm^{-2} at 650°C, showing that good cell performance can be achieved using the thin electrolyte film prepared using the SCS-derived SDC powders (Jiang *et al.*, 2007).

The ternary (Nd-SDC) and quaternary (Pr-Nd-SDC) compositions of doped ceria were processed using citrate–nitrate SC synthesis (Babu and Buri, 2013). Densely sintered compacts with ultrafine structure were achieved by sintering at a lower temperature compared to microcrystalline ceria. Co-doping Pr in ceria increased the oxygen vacancy concentration in the sintered compacts and helped in achieving higher conductivity and lower migration enthalpy than co-doped ceria. It was concluded that a suitable dopant combination that maintains the effective atomic number between 61 and 62 can help to achieve higher conductivity than a binary or ternary system.

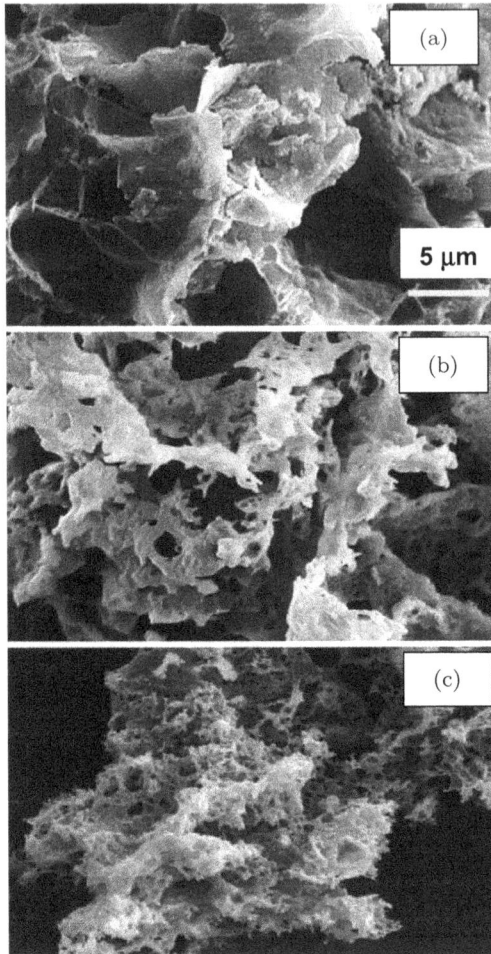

Figure 5.15. Typical SEM images of calcined $Ce_{1-x}Sm_xO_{2-x/2}$ powders prepared using (a) citric acid, (b) alanine, and (c) glycine as the fuels. Magnification: 3000×. Adapted from Chung and Lee (2004).

This can pave the way for developing potential electrolytes with higher ionic conductivity that can operate at lower temperature and can be utilized in IT-SOFCs (Babu and Buri, 2013).

Nanocrystalline $Ce_{0.9}Sm_{0.1}O_{1.95}$ electrolyte powders were prepared by the SC method using urea, citric acid, polyethylene glycol, and glycine fuels (Mangalaraja *et al.*, 2009). The phase-pure

$Ce_{0.9}Sm_{0.1}O_{1.95}$ powders sintered at 1200°C for 6 h exhibited ~98% density. The sintered compacts processed from $Ce_{0.9}Sm_{0.1}O_{1.95}$ powders derived from glycine and citric acid fuels exhibited a maximum Vickers microhardness of 7.94 ± 0.2 and 7.63 ± 0.2 GPa at a load of 20 N. On the other hand, sintered compacts processed from powders derived from urea and polyethylene glycol exhibited maximum fracture toughness values of the order of 3.17 ± 0.3 and 3.06 ± 0.3 MPa m$^{1/2}$, respectively (Mangalaraja *et al.*, 2009). The effect of lattice substitution of 5 mol% Li on the SC-synthesized SDC (20 mol%) (20SDC) has been studied (Basu *et al.*, 2014). Increased ionic conductivity (lattice) at 500°C was observed in $Ce_{0.75}Sm_{0.2}Li_{0.05}O_{1.95}$ compared to that of $Ce_{0.8}Sm_{0.2}O_{1.95}$. Thus, $Ce_{0.75}Sm_{0.2}Li_{0.05}O_{1.95}$ is a potential electrolyte material with respect to the reduced processing temperature as well as the operating temperature of SOFC (Basu *et al.*, 2014). Nanosized $Ce_{0.80}Sm_{0.20}O_{2-\delta}$ (CSO) and $Ce_{0.79}Sm_{0.20}Cu_{0.01}O_{2-\delta}$ (CSCO) were synthesized by the SC method using PVA as fuel. By co-doping CSO with 1 mol% CuO, the electrolyte could be sintered at a lower temperature and it exhibited enhanced mechanical and electrical properties (Dong *et al.*, 2011).

Efforts were made by Lyagaeva *et al.* (2016) to use the SC-synthesized proton-conducting $BaCe_{0.5}Zr_{0.3}Dy_{0.2}O_{3-\delta}$ (BCZD) composition as a reversible SOFC. The BCZD powders synthesized using citrate–nitrate combustion could be sintered at 1450°C to achieve density nearing to theoretical density. The developed BCZD exhibited high stability under high moisture conditions, excellent electrical properties, and chemical and thermal compatibility with cathode and anode materials used in reversible SOFC (R-SOFC) (Lyagaeva *et al.*, 2016). An R-SOFC can operate efficiently both in fuel cell and electrolysis operating modes and the materials used in R-SOFC are those commonly used in SOFC. The grain-boundary resistance of the $BaZr_{0.8}Y_{0.2}O_{3-\delta}$ (BZY20) solid electrolyte prepared using the SC method was reduced markedly compared to BZY20 prepared by conventional SSR (Peng *et al.*, 2014).

As it is difficult to sinter $BaZr_{0.8}Y_{0.2}O_{3-\delta}$ (BZY20) electrolyte, it is challenging to fabricate anode-supported single cells. However, by the partial substitution of Zr with Ni in the BZY perovskite, BZY could be sintered well (Shafi *et al.*, 2015). Phase-pure Ni-doped BZY powders ($BaZr_{0.8-x}Y_{0.2}Ni_xO_{3-\delta}, x = 0.04$) were synthesized using the SC method using citric acid as fuel [90]. $BaZr_{0.76}Y_{0.2}Ni_{0.04}O_{3-\delta}$ (BZYNi04) exhibited sufficient total conductivity; however, the obtained open circuit voltage (OCV) values indicated a lack of significant electronic contribution. The improved sinterability of BZYNiO4 favored the easy fabrication of film and this coupled with the application of an anode functional layer and a suitable cathode, $PrBaCo_2O_{5+\delta}$ (PBCO), yielded an SOFC exhibiting remarkable fuel cell power performance (Shafi *et al.*, 2015). Interestingly, a peak power density of 240 and 428 $mWcm^{-2}$ was obtained using humidified hydrogen and static air as the fuel and oxidant, respectively, at 700°C and 600°C, respectively (Shafi *et al.*, 2015).

The effect of the nature of the precursors, the ratio between the nitrate precursors and the glycine, and the temperature of the calcination step on the microstructure and conductivity of the $BaCe_{0.8}Zr_{0.1}Y_{0.1}O_{3-\delta}$ compound were studied. The highest total conductivity in dry air (above 10^{-2} Scm^{-1} from 700°C) was obtained for a phase-pure sample (Thabet *et al.*, 2018). The electrolyte material with a composition of BCZD synthesized using citrate–nitrate combustion method and sintered at 1450°C for 5 h exhibited 16% shrinkage and 98%density. The as-sintered BCZD exhibited improved electrical properties (19 and 13 $mScm^{-1}$ at 600°C in wet air and wet hydrogen atmospheres, respectively) and acceptable chemical and thermal compatibilities with NiO–BCZD and $La_2NiO_{4+\delta}$–BCZD functional electrodes. An electrochemical cell with a 30-mm thick electrolyte fabricated by tape calendaring method was characterized in SOFC and solid oxide electrolysis cell (SOEC) operation modes. On the basis of comparative analysis of the electrochemical characteristics, it can be inferred that Dy-doped cerate–zirconates as a promising alternative to traditional Y-doped ones. The reversible SOFCs exhibited adequate levels of output characteristics and average ion transport numbers > 0.9 in both the SOFC and SOEC operation modes at 550°C–750°C (Lyagaeva *et al.*, 2016).

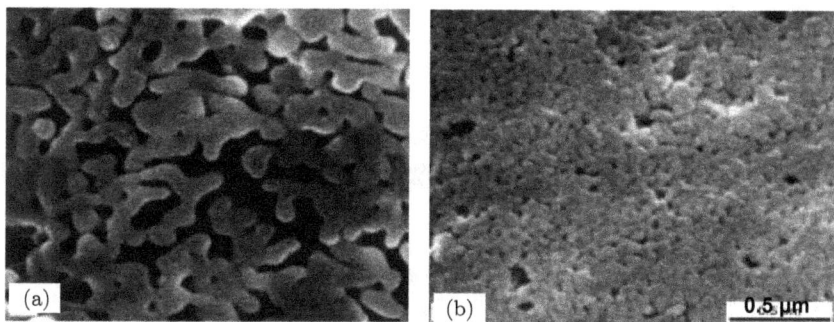

Figure 5.16. Morphology of the combustion product obtained by (a) hot-plate and (b) microwave-induced combustion. Adapted from Park *et al.* (1998).

5.5.4. *SOFC Interconnect*

Park *et al.* (1998) synthesized $LaCrO_3$ powders by microwave-induced combustion of metal nitrate–urea mixture solution. The product obtained using hot plate for ignition exhibited larger agglomerates ($0.12 \, \mu$m) compared to the one obtained by microwave-induced combustion ($0.02 \, \mu$m) (Fig. 5.16). This may be attributed to the nonuniform heating on the hot plate.

The interconnect powder with the composition $La_{0.80}Sr_{0.20}$ $Cr_{0.92}Co_{0.08}O_3$ (LSCrCo) prepared from urea fuel has been evaluated for its rheology and slip-casting behavior (Setz *et al.*, 2011). The suspension for slip casting was prepared using ball milling the synthesized LSCrCo for 24 h, with 3 wt% ammonium polyacrylate and 1 wt% tetramethylammonium hydroxide, respectively. The sintered green discs (1600°C for 4 h) exhibited $> 97\%$ density (Fig. 5.17) (Setz *et al.*, 2011).

In another attempt, the SC-synthesized $La_{0.80}Sr_{0.20}Cr_{0.92}Co_{0.08}$ O_3 interconnect powder prepared from urea as the fuel was dispersed in ethanol with commercial copolymers (Hypermer, KD6) and mixture of binders such as polyvinyl butyral-co-vinyl alcohol-co-vinyl acetate (PVA-PVAc), and two plasticizers, polyethylene glycol (PEG400) and benzyl butyl phthalate (BBP), and tapecast using a tapecaster (Setz *et al.*, 2010). Homogeneous tapes with a controlled thickness ranging from 80 to 230 μm were obtained (Setz *et al.*, 2010).

Figure 5.17.   FESEM image of the fresh fracture surface of the sintered LSCrCo specimen tapecast from optimized suspensions with 17.5 vol% solid loading. Adapted from Setz *et al.* (2011).

5.5.5. *SOFC Single Cells*

The SC method was used to prepare all the nanosized oxides used in IT-SOFCs using glycine as fuel (Ji *et al.*, 2003). A single cell fabricated using SDC as the electrolyte, a composite cathode consisting of LSCF and 50 wt% SDC as the cathode, and Ni-SDC composite powder as the anode and tested at 750°C exhibited a maximum power density of 0.104 Wcm^{-2} and a short circuit current density of 500 mAcm^{-2} (Ji *et al.*, 2003). An anode-supported SOFC consisting of a bi-layered Ni-cermet anode, $Zr_{0.84}Sc_{0.16}O_{1.92}$ electrolyte, and Pt–3% $Zr_{0.84}Y_{0.16}O_{1.92}$ cathode was fabricated using the SC-synthesized powders from glycine fuel (Osinkin *et al.*, 2015). Figure 5.18 shows the photograph of a half cell along with the schematic of the cell. The cathodes and anodes were impregnated with praseodymium and cerium oxides, respectively. The power density of the fuel cell was 0.35 Wcm^{-2} at 900°C and the same fuel cell with impregnated electrodes displayed seven times higher power density (2.4 Wcm^{-2} at 0.6 V) (Osinkin *et al.*, 2015).

Highly crystalline SDC nanoparticles were synthesized using polyacrylamide as the fuel along with the metal nitrates.

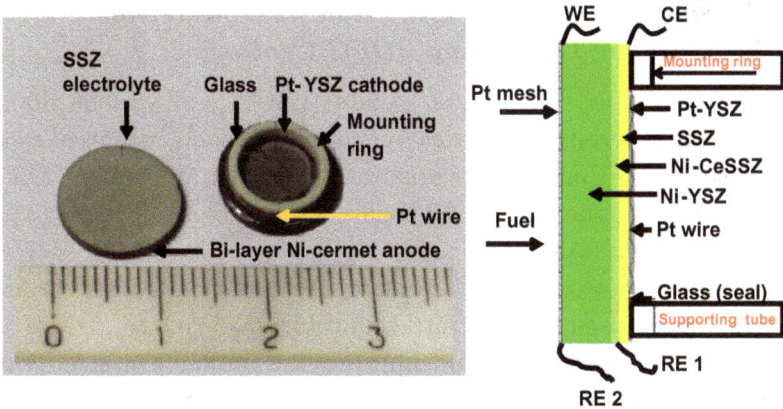

Figure 5.18. Photograph of a half-cell and SOFC (left) and a schematic of SOFC (right). Adapted from Osinkin *et al.* (2015).

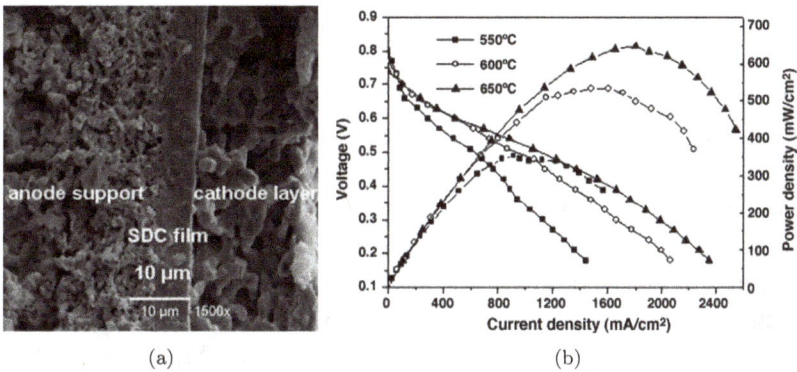

Figure 5.19. (a) SEM cross-sectional image showing the sandwich structure of cell and (b) power output of the cell. Adapted from Dong *et al.* (2009).

Calcined-reactive SDC powders with particle sizes ranging from 0.1 to 1 μm were used to fabricate a dense electrolyte film using the dip-coating method (Fig. 5.19a). Maximum power densities of 353 and 533 mWcm^{-2} were observed, respectively, at 550°C and 600°C, which was ascribed to low cathode polarization resistance (Fig. 5.19b). Thus, it was concluded that polymer hydrogel-assisted SC method is a facile process for synthesizing highly reactive nanoparticles for many applications, including SOFC (Dong *et al.*, 2009).

5.5.6. *Solution Combustion-Synthesized Materials Used for Developing SOFC Coatings*

Sr- and Mg-doped lanthanum gallate (LSGM), of various dopant concentrations prepared by urea/nitrate SC method, was used as the target source for pulsed-laser ablation to deposit amorphous LSGM films, which on annealing at 973 K gave a single-phase cubic structure (Mathews *et al.*, 2000).

In recent years, in-lieu of ceramic interconnects ferritic stainless steel is being used as the interconnector material to reduce the cost of the raw materials and improve the mechanical strength of supported thin film fuel cell membranes. However, there is a need to suppress chromium vaporization that results in the formation of oxidation scales on ferritic stainless steel that leads to the deterioration of the cell performance. Combustion-synthesized lanthanum strontium manganese oxide (LSM: $La_{0.65}Sr_{0.3}MnO_3$), lanthanum strontium cobalt iron oxide (LSCF: $La_{0.6}Sr_{0.4}Co_{0.8}Fe_{0.2}O_3$), and manganese cobalt oxide (MCO: $MnCo_2O$) were slurry sprayed from the SC-synthesized submicron powders, which reduced the chromium vaporization by a factor of 2–3 compared to uncoated interconnect (Kurokawa *et al.*, 2007).

Mn-Co oxide coating has been developed by the sputtering technique using a target fabricated by uniaxial pressing of the SC-synthesized powder (Kumar *et al.*, 2016). The Co-Mn oxide powder (Fig. 5.20) corresponding to the composition $Mn_{1.5}Co_{1.5}O_4$ was prepared using cobalt and manganese salts in 1:1 mole ratio and 13.6 g of glycine (Kumar *et al.*, 2016). The resulting ceramic powder was calcined at 800°C for 2 h to remove traces of carbon impurities followed by ball milling for 2 h for breaking the soft agglomerates and pelletized in a hydraulic press by adding 6% PVA to the powder. The prepared pellets were calcined in air at 800°C for 2 h and later sintered at 1150°C for 2 h and used as the sputtering target. SS-430 coupon with and without spinel coating was subjected to oxidation at 800°C for a period of 72 h in flowing air and the coated coupon exhibited better oxidation resistance compared to the uncoated substrate (Fig. 5.21).

Figure 5.20. FESEM image of SC-synthesized $Mn_{1.5}Co_{1.5}O_4$ powder. Adapted from Kumar *et al.* (2016).

(a)

(b)

Figure 5.21. Oxidized SS-430 coupons: (a) $Mn_{1.5}Co_{1.5}O_4$ sputter coated and (b) bare coupon.

Plasma-sprayable grade-flowable GDC powder was prepared by fuel-deficient combustion reaction using glycine as fuel (Shri Prakash *et al.*, 2017). Electrolyte coatings fabricated from the synthesized plasma spray grade powders exhibited superior inter splat adhesion and conductivity ($\sim$0.02 Scm^{-1} at 600°C) (Fig. 5.22). The obtained dense nature of the coating was assigned to the complete melting of the porous GDC particles in the plasma flame. Powders prepared with stoichiometric fuel were found to be suitable for tape casting process (Shri Prakash *et al.*, 2017).

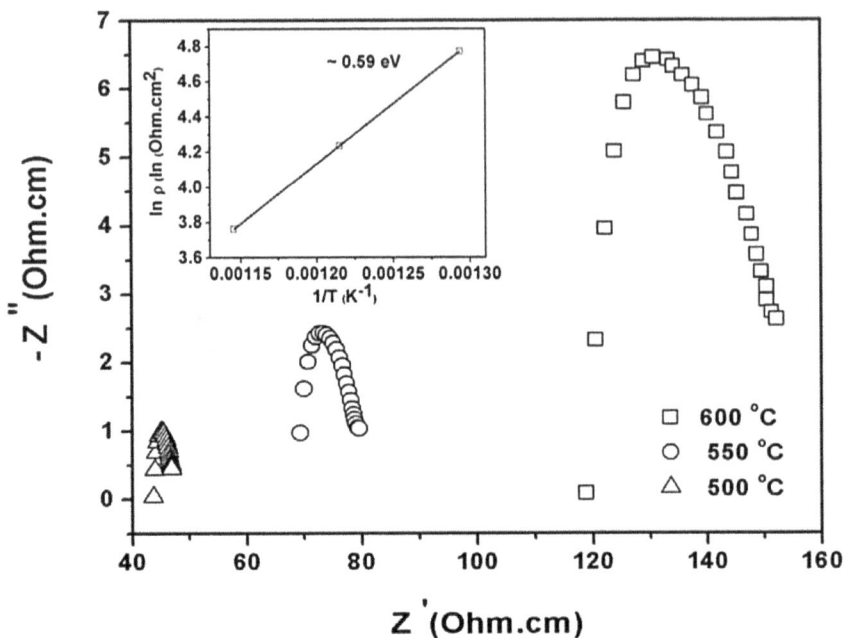

Figure 5.22. Impedance plot of plasma-sprayed GDC at different temperatures. Adapted from Shri Prakash *et al.* (2017).

An all-perovskite-based IT fuel cell was fabricated from materials synthesized using glycine-nitrate process and modified Pechini synthesis routes (Pine *et al.*, 2007). Yttrium-doped strontium titanate (SYT) was chosen as the conductive ceramic anode component to avoid problems associated with Ni-based anodes. $La_{0.8}Sr_{0.2}Ga_{0.8}Mg_{0.1}Co_{0.1}O_3$−ES SOFCs with composite ceramic anode and cathode exhibited a relatively high power density of $0.246\,Wcm^{-2}$ at 800°C at 0.5 V (Pine *et al.*, 2007).

5.6. Conclusion

The SC method is a versatile, simple, and cost-effective process for the fabrication of SOFC components. The SCS has been successfully explored for the fabrication of micron and nanosized SOFC electrode and electrolyte materials. Researchers have explored various fuels and oxidizer/fuel ratio to synthesize SOFC materials with improved properties compared to those prepared from other chemical routes. The synthesized materials have been successfully used for the fabrication of complete SOFC single cells. Also, the synthesized powders have been used for the fabrication of coatings by (1) sputtering, (2) spin coating, (3) wet powder spraying, (4) tapecasting, (5) tape calendaring, and (6) plasma spraying. Studies have also shown that citric acid is a better fuel for synthesizing nanosized oxides for SOFC application. It is also gratifying to note that one can prepare powders suitable for tapecasting and plasma spraying, wherein the tapecasting process requires submicron-sized powders and plasma spraying process requires flowable micron-sized particles. The SCS-synthesized nanoparticles are of high-quality materials suitable for SOFC application. However, the challenge is to scale-up the SC process for the preparation of large quantities of SOFC powders.

Acknowledgments

The authors acknowledge the encouragement and support received from Director, CSIR-NAL and Head, Surface Engineering Division, CSIR-NAL.

References

Internet

1. http://www.owlnet.rice.edu/~tl5/EnergyIEEJ.pdf accessed on 30th January 2019.
2. http://www.eai.in/ref/ct/fc/fuel_cells.html accessed on 30th January 2019.
3. http://www.ausairpower.net/SP/DT-AIP-SSK-Dec-2010.pdf.
4. http://www.11ecpss.betterbtr.com/pdf%20folder%20wednesday/W15-Burke_PowerSources_Final.pdf.
5. https://www.techrepublic.com/article/ges-new-fuel-cell-technology-is-a-game29changer/accessed on 30th January 2019.

Journals & Books

Accardo, G., Frattini, D., Ham, H.C. & Yoon, S.P. (2019). Direct addition of lithium and cobalt precursors to $Ce_{0.8}Gd_{0.2}O_{1.95}$ electrolytes to improve microstructural and electrochemical properties in IT-SOFC at lower sintering temperature. *Ceram. Int., 45*, 9348–9358.

Ajamein, H., & Haghighi, M. (2015). Effect of sorbitol/oxidizer ratio on microwave assisted solution combustion synthesis of copper based nanocatalyst for fuel cell grade hydrogen production. *Iranian J. Hyd. & Fuel Cell, 4*, 227–240.

Ainirad, A., Kashani Motlagh, M. M., & Maghsoudipoor, A. (2011). A systematic study on the synthesis of Ca, Gd codoped cerium oxide by combustion method. *J. Alloys Compd., 509*, 1505–1510.

Akbari-Fakhrabadi, A., Avila, R. E., Carrasco, H. E., Ananthakumar, S., & Mangalaraja, R.V. (2012). Combustion synthesis of $NiO\text{-}Ce_{0.9}Gd_{0.1}O_{1.95}$ nanocomposite anode and its electrical characteristics of semi-cell configured SOFC assembly. *J. Alloys Compd., 541*, 1–5.

Akbari-Fakhrabadi, A., Mangalaraja, R. V., Sanhueza, F. A., Avila, R. E., Ananthakumar, S., & Chan, S. H. (2012). Nanostructured $Gd\text{-}CeO_2$ electrolyte for solid oxide fuel cell by aqueous tape casting. *J. Power Sources, 218*, 307–312.

Aliotta, C., Liotta, L. F., La Parola, V., Martorana, A., Muccillo, E. N. S., Muccillo, R., . . . Deganello, F. (2016). Ceria-based electrolytes prepared by solution combustion synthesis: The role of fuel on the materials properties. *Appl. Catal. B: Environ., 197*, 14–22.

Aruna, S. T. (1998). Combustion synthesis and properties of nanosized oxides: Studies on Solid oxide fuel cell materials, *Ph.D Thesis Dissertation, Indian Institute of Science*, India.

Aruna, S. T., & Mukasyan, A. S. (2008). Combustion synthesis and nanomaterials. *Curr. Opin. Solid State Mater. Sci., 12*(3–4), 44–50.

Aruna, S. T., Muthuraman, M., & Patil, K. C. (1997). Combustion synthesis and properties of strontium substituted lanthanum manganites $La_{1-x}Sr_xMnO_3$ (0 < x < 0.3). *J. Mater. Chem., 7*(12), 2499–2503.

Aruna, S. T., Muthuraman, M., & Patil, K. C. (2000). Studies on combustion synthesized $LaMnO_3\text{-}LaCoO_3$ solid solutions. *Mater. Res. Bull., 35*(2), 289–296.

Babu, A. S., & Bauri, R. (2013). Rare earth co-doped nanocrystalline ceria electrolytes for intermediate temperature solid oxide fuel cells (IT-SOFC). *ECS Trans., 57*(1), 1115–1123.

Bansal, N. P., & Zhong, Z. (2006). Combustion synthesis of $Sm_{0.5}Sr_{0.5}CoO_{3-x}$ and $La_{0.6}Sr_{0.4}CoO_{3-x}$ nanopowders for solid oxide fuel cell cathodes. *J. Power Sources, 158*(1), 148–153.

Bastidas, D. M., Tao, S., & Irvine, J. T. S. (2006). A symmetrical solid oxide fuel cell demonstrating redox stable perovskite electrodes. *J. Mater. Chem., 16*(17), 1603–1605.

Basu, R. N., Tietz, F., Wessel, E., Buchkremer, H. P., & Stöver, D. (2004). Microstructure and electrical conductivity of $LaNi_{0.6}Fe_{0.4}O_3$ prepared by combustion synthesis routes. *Mater. Res. Bull., 39*(9), 1335–1345.

Basu, S., Khamrui, S., & Bandyopadhyay, N. R. (2014). Sintering and electrical properties of $Ce_{0.75}Sm_{0.2}Li_{0.05}O_{1.95}$. *Int. J. Hydrogen Energy, 39*(30), 17429–17433.

Biswas, M. K., & Sadanala, C. (2013). Electrolyte materials for solid oxide fuel cell. *J. Powder Metall. Min., 2*, 117.

Buckingham, J., Hodge, J. C. & Hardy, T. (2008). "Submarine power and propulsion- application of technology to deliver customer benefit." *Presented in Submar. power Propuls. Technol. Dev. UDT Europe*, Glasgow.

Cai, R., Zhou, W., & Shao, Z. (2010). Cellulose-assisted combustion synthesis of functional materials for energy storage or conversion. In *Combustion Synthesis: Novel Routes to Novel Materials* (pp. 72–82).

Caronna, T., Fontana, F., Natali Sora, I., Pelosato, R., & Viganò, L. (2010). Chemical compatibility of Sr-doped La_2CuO_4 cathode material with LSGM solid oxide electrolyte. *Solid State Ionics, 181*(29–30), 1355–1358.

Chanquía, C. M., Mogni, L., Troiani, H. E., & Caneiro, A. (2014). Highly active $La_{0.4}Sr_{0.6}Co_{0.8}Fe_{0.2}O_{3-\delta}$ nanocatalyst for oxygen reduction in intermediate temperature-solid oxide fuel cells. *J. Power Sources, 270*, 457–467.

Chinarro, E., Jurado, J. R., & Colomer, M. T. (2007). Synthesis of ceria-based electrolyte nanometric powders by urea-combustion technique. *J. Eur. Ceram. Soc., 27*(13–15), 3619–3623.

Chung, D. Y., & Lee, E. H. (2004). Microwave-induced combustion synthesis of $Ce_{1-x}Sm_xO_{2-x/2}$ powder and its characterization. *J. Alloys Compd., 374*(1–2), 69–73.

da Conceição, L., Ribeiro, N. F. P., Furtado, J. G. M., & Souza, M. M. V. M. (2009). Effect of propellant on the combustion synthesized Sr-doped $LaMnO_3$ powders. *Ceram. Int., 35*(4), 1683–1687.

Deganello, F., Esposito, V., Traversa, E., & Miyayama, M. (2005). Electrode performance of nanostructured $La_{1-a}Sr_aCo_{1-b}Fe_bO_{3-x}$ on a $Ce_{0.8}Sm_{0.2}O_2$ electrolyte prepared by citrate nitrate auto-combustion. *ECS Trans., 1*(7), 219–232.

Dong, D., Chai, Z., Li, C. Z., & Wang, H. (2009). Polymer hydrogel assisted combustion synthesis of highly crystalline ceramic nanoparticles for SOFC electrolyte films. *Mater. Chem. Phys., 118*(1), 148–152.

Dong, D., Gao, J., Liu, M., Chu, G., Diwu, J., Liu, X., ... Meng, G. (2008). Preparation of $Pr_{0.35}Nd_{0.35}Sr_{0.3}MnO_{3-\delta}$/YSZ composite cathode powders for tubular solid oxide fuel cells by microwave-induced monomer gelation and gel combustion synthesis process. *J. Power Sources, 175*(1), 436–440.

Dong, D., Li, C. Z., & Wang, H. (2010). Combustion synthesis of ceramic nanoparticles for solid oxide fuel cells. *Asia-Pac. J. Chem. Eng., 5*(4), 593–598.

Dong, Y., Hampshire, S., Zhou, J. E., Dong, X., Lin, B., & Meng, G. (2011a). Combustion synthesis and characterization of Cu-Sm co-doped CeO_2 electrolytes. *J. Eur. Ceram. Soc., 31*(13), 2365–2376.

Dong, Y., Hampshire, S., Zhou, J. E., & Meng, G. (2011b). Synthesis and sintering of Gd-doped CeO_2 electrolytes with and without 1 at. % CuO doping for solid oxide fuel cell applications. *Int. J. Hydrogen Energy, 36*(8), 5054–5066.

Dutta, A., Götz, H., Ghosh, S., & Basu, R. N. (2009a). Combustion synthesis of $La_{0.6}Sr_{0.4}Co_{0.98}Ni_{0.02}O_3$ cathode and evaluation of its electrical and electrochemical properties for IT-SOFC. *ECS Trans., 25*(2 PART 3), 2657–2666.

Dutta, A., Mukhopadhyay, J., & Basu, R. N. (2009b). Combustion synthesis and characterization of LSCF-based materials as cathode of intermediate temperature solid oxide fuel cells. *J. Eur. Ceram. Soc.*, *29*(10), 2003–2011.

Dutta, A., Patra, S., Bedekar, V., Tyagi, A. K., & Basu, R. N. (2009c). Nanocrystalline gadolinium doped ceria: Combustion synthesis and electrical characterization. *J. Nanosci. Nanotechnol.*, *9*(5), 3075–3083.

Gottmann, M. (2005). Solid oxide regenerative fuel cell for airplane power generation and storage. US 6854688 B2.

Jadhav, L. D., Chourashiya, M. G., Subhedar, K. M., Tyagi, A. K., & Patil, J. Y. (2009). Synthesis of nanocrystalline Gd doped ceria by combustion technique. *J. Alloys Compd.*, *470*(1–2), 383–386.

Jamale, A. P., Bhosale, C. H., & Jadhav, L. D. (2015). Electrochemical behavior of LSCF/GDC interface in symmetric cell: An application in solid oxide fuel cells. *J. Alloys Compd.*, *623*, 136–139.

Ji, Y., Liu, J., He, T., Cong, L., Wang, J., & Su, W. (2003). Single intermedium-temperature SOFC prepared by glycine-nitrate process. *J. Alloys Compd.*, *353*(1–2), 257–262.

Jiang, C., Ma, J., Liu, X., & Meng, G. (2007). Electrochemical performance of a solid oxide fuel cell based on $Ce_{0.8}Sm_{0.2}O_{2-\delta}$ electrolyte synthesized by a polymer assisted combustion method. *J. Power Sources, 165*(1), 134–137.

Jo, S. H., Muralidharan, P., & Kim, D. K. (2009). Enhancement of electrochemical performance and thermal compatibility of $GdBaCo_{2/3}Fe_{2/3}Cu_{2/3}O_{5+\delta}$ cathode on $Ce_{1.9}Gd_{0.1}O_{1.95}$ electrolyte for IT-SOFCs. *Electrochem. Commun.*, *11*(11), 2085–2088.

Jurado, J. R. (2001). Present several items on ceria-based ceramic electrolytes: Synthesis, additive effects, reactivity and electrochemical behaviour. *J. Mater. Sci.*, *36*(5), 1133–1139.

Khandale, A. P., Punde, J. D., & Bhoga, S. S. (2013). Electrochemical performance of strontium-doped neodymium nickelate mixed ionic-electronic conductor for intermediate temperature solid oxide fuel cells. *J. Solid State Electrochem.*, *17*(3), 617–626.

Kiminami, R. H. G. A., & Morelli. M. R. (2004). Microwave-assisted combustion synthesis of $LaCrO_3$ nanopowders. *J. Metastable Nanocrystal. Mater.*, *22*, 91–96.

Kumar, S. A., Kuppusami, P., & Vengatesh, P. (2018). Auto-combustion synthesis and electrochemical studies of $La_{0.6}Sr_{0.4}Co_{0.2}Fe_{0.8}O_{3-\delta}-Ce_{0.8}Sm_{0.1}Gd_{0.1}O_{1.90}$ nanocomposite cathode for intermediate temperature solid oxide fuel cells. *Ceram. Int.*, *44*(17), 21188–21196.

Kumar, S. S., Nalluri, A., Anandan, C., Prakash, B. S., & Aruna, S. T. (2016). Deposition and evaluation of Mn-Co oxide protective sputtered coating on SOFC interconnects and current collectors. *J. Electrochem. Soc.*, *163*(8), F905–F912.

Kurokawa, H., Jacobson, C. P., DeJonghe, L. C., & Visco, S. J. (2007). Chromium vaporization of bare and of coated iron-chromium alloys at 1073 K. *Solid State Ionics*, *178*(3–4), 287–296.

Larminie, J., Dicks, A. (2003). *Fuel Cell Systems Explained* (2nd ed.). England: John Wiley & Sons Ltd.

Lee, S. J., Kim, D. S., Muralidharan, P., Jo, S. H., & Kim, D. K. (2011). Improved electrochemical performance and thermal compatibility of Fe- and Cu-doped $SmBaCo_2O_{5+\delta}$-$Ce_{0.9}Gd_{0.1}O_{1.95}$ composite cathode for intermediate temperature solid oxide fuel cells. *J. Power Sources, 196*(6), 3095–3098.

Li, F. T., Ran, J., Jaroniec, M., & Qiao, S. Z. (2015). Solution combustion synthesis of metal oxide nanomaterials for energy storage and conversion. *Nanoscale, 7*(42), 17590–17610.

Liu, B., Zhang, Y., & Zhang, L. (2008). Characteristics of $Ba_{0.5}Sr_{0.5}Co_{0.8}Fe_{0.2}O_{3-\delta}$-$La_{0.9}Sr_{0.1}Ga_{0.8}Mg_{0.2}O_{3-\delta}$ composite cathode for solid oxide fuel cell. *J. Power Sources, 175*(1), 189–195.

Lyagaeva, J., Danilov, N., Vdovin, G., Bu, J., Medvedev, D., Demin, A., ... Tsiakaras, P. (2016). A new Dy-doped $BaCeO_3$-$BaZrO_3$ proton-conducting material as a promising electrolyte for reversible solid oxide fuel cells. *J. Mater. Chem. A, 4*(40), 15390–15399.

Magnone, E., Miyayama, M., & Traversa, E. (2006). $Pr_{0.8}Sr_{0.2}Co_{1-x}Fe_xO_3$ nanocrystalline powders: Electrochemical characterization for solid oxide fuel cell cathodes. *ECS Trans., 3*(9), 163–171.

Magnone, E., Miyayama, M., & Traversa, E. (2010). Electrochemical impedance spectroscopy analysis of $Pr_{0.8}Sr_{0.2}Co_{0.5}Fe_{0.5}O_{3-\delta}$ as cathode material for intermediate temperature solid oxide fuel cells. *J. Electrochem. Soc., 157*(3), B357–364.

Magnone, E., Traversa, E., & Miyayama, M. (2009). Nano-sized $Pr_{0.8}Sr_{0.2}Co_{1-x}Fe_xO_3$ powders prepared by single-step combustion synthesis for solid oxide fuel cell cathodes. *J. Electroceram., 2*, 122–135.

Magrasó, A., Calleja, A., Capdevila, X. G., & Espiell, F. (2004). Synthesis of Gd-doped $BaPrO_3$ nanopowders. *Solid State Ionics, 166*(3–4), 359–364.

Mahato, N., Banerjee, A., Gupta, A., Omar, S., & Balani, K. (2015). Progress in material selection for solid oxide fuel cell technology: A review. *Prog. Mater. Sci., 72*, 141–337.

Mangalaraja, R. V., Ananthakumar, S., Uma, K., Jiménez, R. M., López, M., & Camurri, C. P. (2009). Microhardness and fracture toughness of $Ce_{0.9}Gd_{0.1}O_{1.95}$ for manufacturing solid oxide electrolytes. *Mater. Sci. Eng., A, 517*(1–2), 91–96.

Mather, G. C., Figueiredo, F. M., Jurado, J. R., & Frade, J. R. (2004). Electrochemical behaviour of Ni-cermet anodes containing a proton-conducting ceramic phase on YSZ substrate. *Electrochim. Acta, 49*(16), 2601–2612.

Mathews, T., Manoravi, P., Antony, M. P., Sellar, J. R., & Muddle, B. C. (2000). Fabrication of $La_{1-x}Sr_xGa_{1-y}Mg_yO_{3-(x+y)/2}$ thin films by pulsed laser ablation. *Solid State Ionics, 135*(1–4), 397–402.

Moreno, B., Fernández-González, R., Jurado, J. R., Makradi, A., Nuñez, P., & Chinarro, E. (2014). Fabrication and characterization of ceria-based buffer layers for solid oxide fuel cells. *Int. J. Hydrogen Energy, 39*(10), 5433–5439.

Nielsen, J., Persson, Å. H., Thanh Muhl, T., & Brodersen, K. (2018). Towards high power density metal supported solid oxide fuel cell for mobile applications. *J. Electrochem. Soc., 165*(2), F90–F96.

Osinkin, D. A., Bogdanovich, N. M., Beresnev, S. M., & Zhuravlev, V. D. (2015). High-performance anode-supported solid oxide fuel cell with impregnated electrodes. *J. Power Sources, 288*, 20–25.

Park, H. K., Han, Y. S., Kim, D. K., & Kim, C. H. (1998). Synthesis of $LaCrO_3$ powders by microwave induced combustion of metal nitrate-urea mixture solution. *J. Mater. Sci. Lett., 17*, 785–787.

Patil, K. C., Aruna, S. T., & Ekambaram, S. (1997). Combustion synthesis. *Curr. Opin. Solid State Mater. Sci., 2*(2), 158–165.

Patil, K. C., Aruna, S. T., & Mimani, T. (2002). Combustion synthesis: An update. *Curr. Opin. Solid State Mater. Sci., 6*(6), 507–512.

Patil, K. C., Hegde, M. S., Tanu Rattan., & Aruna, S. T. (2008). *Chemistry of Nanocrystalline Oxide Materials: Combustion Synthesis, Properties and Applications*. Singapore: World Scientific.

Peng, C., Huang, T., & Zheng, Y. (2014). Comparative study on preparation and properties of $BaZr_{0.8}Y_{0.2}O_{3-\delta}$ solid electrolyte. *Key Eng. Mater., 591*, 240–244.

Pine, T. S., Lu, X., Do, A. T. V., Mumm, D. R., & Brouwer, J. (2007). Operation of an LSGMC electrolyte-supported SOFC with composite ceramic anode and cathode. *Electrochem. Solid-State Lett., 10*(10), B183–B185.

Rahaman, M. N. (2006). *Ceramic Processing*. Boca Raton: CRC Press.

Ringuedé, A., Bronine, D., & Frade, J. R. (2002). $Ni_{1-x}Co_x$/YSZ cermet anodes for solid oxide fuel cells. *Electrochim. Acta, 48*(4), 437–442.

Ringuede, A., Labrincha, J. A., & Frade, J. R. (2001). A combustion synthesis method to obtain alternative cermet materials for SOFC anodes. *Solid State Ionics, 141–142*, 549–557.

Saha, S., Ghanawat, S. J., & Purohit, R. D. (2006). Solution combustion synthesis of nano particle $La_{0.9}Sr_{0.1}MnO_3$ powder by a unique oxidant-fuel combination and its characterization. *J. Mater. Sci., 41*(7), 1939–1943.

Setz, L. F. G., de Mello-Castanho, S. R. H., Colomer, M. T., & Moreno, R. (2009). Surface behaviour and stability of strontium and cobalt doped-lanthanum chromite powders in water, *Solid State Ionics, 180*(1), 71–75.

Setz, L. F. G., Santacruz, I., Colomer, M. T., Mello-Castanho, S. R. H., & Moreno, R. (2010). Tape casting of strontium and cobalt doped lanthanum chromite suspensions. *J. Eur. Ceram. Soc., 30*(14), 2897–2903.

Setz, L. F. G., Santacruz, I., Colomer, M. T., Mello-Castanho, S. R. H., & Moreno, R. (2011). Fabrication of Sr- and co-doped lanthanum chromite interconnectors for SOFC. *Mater. Res. Bull., 46* (7), 983–986.

Shafi, S. P., Bi, L., Boulfrad, S., & Traversa, E. (2015). Y and Ni Co-doped $BaZrO_3$ as a proton-conducting solid oxide fuel cell electrolyte exhibiting superior power performance. *J. Electrochem. Soc., 162*(14), F1498–1F503.

Shri Prakash, B., Balaji, N., Senthil Kumar, S., & Aruna S. T. (2016a). Microstructure and polarization studies on interlayer free

$La_{0.65}Sr_{0.3}Co_{0.2}Fe_{0.8}O_{3-d}$cathodes fabricated on yttria stabilized zirconia by solution precursor plasma spraying. *Fuel Cells, 16*, 617–627.

Shri Prakash, B., Balaji, N., Senthil Kumar, S., & Aruna, S. T. (2016b). Properties of nano-structured Ni/YSZ anodes fabricated from plasma sprayable NiO/YSZ powder prepared by single step solution combustion method. *Applied Surface Science, 389*, 983–989.

Shri Prakash, B., Parthasarathi, B., Senthil Kumar, S., & Aruna, S. T. (2017). Microstructure and electrical properties of plasma sprayed $Gd_{0.15}Ce_{0.85}O_{2-\delta}$ coatings from solution combustion synthesized flowable powders. *J. Eur. Ceram. Soc., 37*, 271–279.

Shri Prakash, B., Senthil Kumar, S., & Aruna, S. T. (2014). Properties and development of Ni/YSZ as an anode material in solid oxide fuel cell: A Review. *Renew. Sust. Energ. Rev., 36*, 149–179.

Shri Prakash, B., William Grips, V. K., & Aruna, S.T. (2014). A single step solution combustion approach for preparing gadolinia doped ceria solid oxide fuel cell electrolyte material suitable for wet powder and plasma spraying processes. *J. Power Sources, 214,* 358–364.

Singh, N. K., Singh, P., Singh, M. K., Kumar, D., & Prakash, O. (2011). Auto-combustion synthesis and properties of $Ce_{0.85}Gd_{0.15}O_{1.925}$ for intermediate temperature solid oxide fuel cells electrolyte. *Solid State Ionics, 192*(1), 431–434.

Singhal, S. C., & Kendall, K. (2003). *High-Temperature Solid Oxide Fuel Cells: Fundamentals, Design and Applications* (1st ed.). Elsevier.

Stevenson, J. W., Armstrong, T. R., McCready, D. E., Pederson, L. R., & Weber, W. J. (1997). Processing and electrical properties of alkaline earth-doped lanthanum gallate. *J. Electrochem. Soc., 144*(10), 3613–3620.

Sun, C., Rob Hui, S., & Roller, J. (2010). Cathode materials for solid oxide fuel cells: A review. *J. Solid State Electr., 14*(7), 1125–1144.

Sun, C., & Stimming, U. (2007). Recent anode advances in solid oxide fuel cells. *J. Power Sources, 171*, 247–260.

Thabet, K., Devisse, M., Quarez, E., Joubert, O., & Le Gal La Salle, A. (2018). Influence of the autocombustion synthesis conditions and the calcination temperature on the microstructure and electrochemical properties of $BaCe_{0.8}Zr_{0.1}Y_{0.1}O_{3-\delta}$ electrolyte material. *Solid State Ionics, 325*, 48–56.

Varma, A., Mukasyan, A. S., Rogachev, A. S., & Manukyan, K. V. (2016). Solution combustion synthesis of nanoscale materials. *Chem. Rev., 116*(23), 14493–14586.

Vijaya Lakshmi, V., & Bauri, R. (2011a). Phase formation and ionic conductivity studies on ytterbia co-doped scandia stabilized zirconia ($0.9ZrO_2$-$0.09Sc_2O_3$-$0.01Yb_2O_3$) electrolyte for SOFCs. *Solid State Sci., 13*(8), 1520–1525.

Vijaya Lakshmi, V., Bauri, R., Gandhi, A. S., & Paul, S. (2011b). Synthesis and characterization of nanocrystalline ScSZ electrolyte for SOFCs. *Int. J. Hydrogen Energy, 36*(22), 14936–14942.

Wu, J., & Gupta, A. (2010). Recent development of SOFC metallic interconnect. *J. Mater. Sci. Technol., 26*(4), 293–305.

Xu, H., Yan, H., & Chen, Z. (2008). Preparation and properties of Y^{3+} and Ca^{2+} co-doped ceria electrolyte materials for ITSOFC. *Solid State Sci., 10*(9), 1179–1184.

Xue, J., Shen, Y., Zhou, Q., He, T., & Han, Y. (2010). Combustion synthesis and properties of highly phase-pure perovskite electrolyte co-doped $La_{0.9}Sr_{0.1}Ga_{0.8}Mg_{0.2}O_{2.85}$ for IT-SOFCs. *Int. J. Hydrogen Energy, 35*(1), 294–300.

Yang, S. H., Kim, K. H., Yoon, H. H., Kim, W. J., & Choi, H. W. (2011). Comparison of combustion and solid-state reaction methods for the fabrication of SOFC LSM cathodes. *Mol. Cryst. Liq. Cryst., 539*, 390–397.

Yildiz, O., Soydan, A. M., Akel, M., Ipçizade, E. F., & Ata, A. (2015). Characterization of MOy-$M_xCe_{1-x}O_{2-\delta}$ (M: Co, Ni & Cu) nano powders and anode materials for low and intermediate temperature solid oxide fuel cells. *Int. J. Hydrogen Energy, 40*(40), 14085–14094.

Zarkov, A., Stanulis, A., Sakaliuniene, J., Butkute, S., Abakeviciene, B., Salkus, T., ... Kareiva, A. (2015). On the synthesis of yttria-stabilized zirconia: A comparative study. *J. Sol-Gel Sci. Technol., 76*(2), 309–319.

Zhu, C., Nobuta, A., Ju, Y. W., Ishihara, T., & Akiyama, T. (2013). Solution combustion synthesis of $Ce_{0.6}Mn_{0.3}Fe_{0.1}O_2$ for anode of SOFC using $LaGaO_3$-based oxide electrolyte. *Int. J. Hydrogen Energy, 38*(30), 13419–13426.

Chapter 6

Green SCS: Hydrogen Production, Further Applications, Conclusion, and Path Forward

Sergio L. González-Cortés

*Inorganic Chemistry Laboratory, Department of Chemistry,
University of Oxford, South Parks Road, Oxford,
OX1 3QR, United Kingdom
sergio.gonzalez-cortes@chem.ox.ac.uk;
slgoncor@gmail.com*

6.1. Introduction

Hydrogen is considered an important energy carrier and alternative source for sustainable and clean energy systems in the stationary power, transportation, industrial, and residential sectors. This can be related to a series of factors such as its flexibility associated with its production and use, its relatively high abundance in the Earth (Navlani-García *et al.*, 2018), and its high energy content relative to methanol and conventional fuels (Dincer and Acar, 2015) as shown in Fig. 6.1, and also to the production of only by-product water — no carbon dioxide (CO_2) — on the combustion reaction. Evidently, additional measures to decarbonize hydrogen-production technologies need to be in place to mitigate the carbon emissions. Otherwise, the CO_2 emissions are simply moving upstream from the end-use side to the hydrogen-production sector.

Hydrogen can be stored as compressed gas, cryogenic liquid, solid hydride, hydrocarbons, and even as waste plastics; it can also be used as a fuel either for direct combustion in an internal combustion engine or in a fuel cell device. Furthermore, hydrogen is not a primary energy

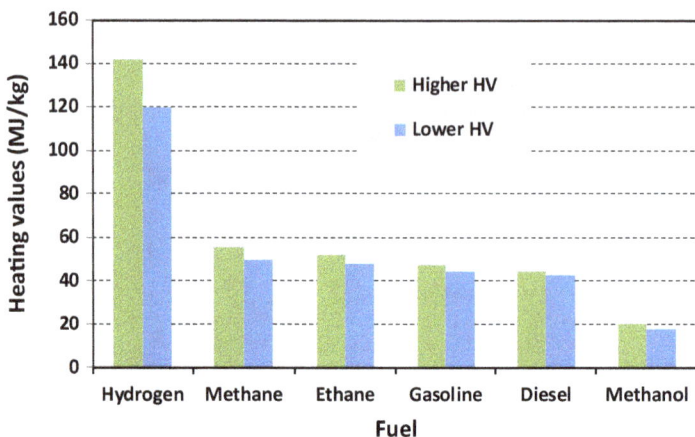

Figure 6.1. Higher and lower heating values for hydrogen- and carbon-containing fuels. Data adapted from Dincer and Acar (2015).

source as coal, gas, or oil. It is indeed designated as an energy carrier (or energy vector), which is produced from another source and then transported and stored for future use (Edwards *et al.*, 2008). A wide variety of processes are available for hydrogen production, which according to the raw materials used, could be divided into two major categories: conventional (fossil fuels) and renewable technologies as shown in Fig. 6.2.

The first category includes the processing of fossil fuels by reforming and pyrolysis. In hydrocarbon reforming, the participating processes are steam reforming (SR), partial oxidation (PO), and autothermal steam reforming (ATR). The second category (i.e., renewable technologies) involves the methods that produce hydrogen from either water or biomass (renewable sources). The hydrogen production from water-splitting processes includes electrolysis, thermolysis, and photoelectrolysis. The utilization of biomass as a feedstock involves two general subcategories, namely thermochemical and biological processes. Thermochemical technology mainly involves pyrolysis, gasification, combustion, and liquefaction, whereas the major biological processes are direct and indirect bio-photolysis, dark fermentation, photofermentation, and sequential dark (Holladay *et al.*, 2009; Rodriguez-Sulbaran *et al.*, 2018).

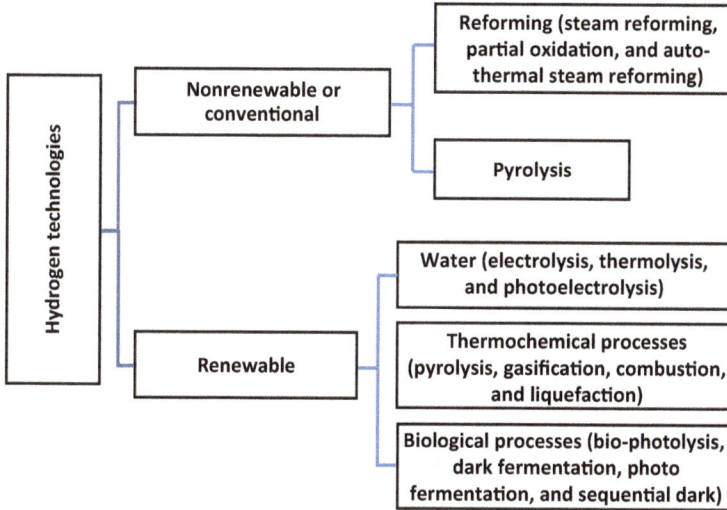

Figure 6.2. Schematic representation of different technologies to produce hydrogen.

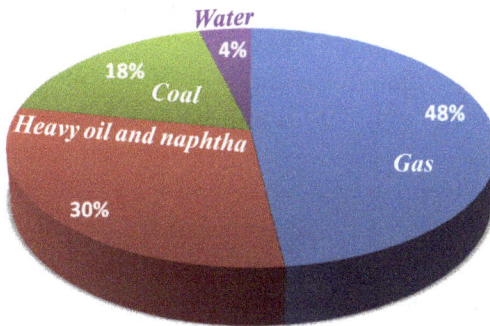

Figure 6.3. Commercial production of hydrogen from various sources. Adapted from Kothari *et al.* (2008).

The commercial production of hydrogen is from four main sources: natural gas (48%), liquid oil (30%), coal (18%), and electrolysis (4%) as shown in Fig. 6.3 (Kothari *et al.*, 2008). Fossil fuels are the main sources of hydrogen at industrial level, and SR is one of the most widespread and least expensive processes for hydrogen production (Kalamaras and Efstathiou, 2013).

However, a number of factors such as energy security, limited reserves of fossil fuels, increased environmental concerns about greenhouse gas (GHG) emissions, and global climate change will gradually decrease the fossil fuel production and consumption over time (WEO, 2018). It also motivates the research for affordable, reliable, and renewable fuel energy sources that could greatly reduce the GHG emissions and their adverse effects on global warming. It is envisaged that catalysis will continue playing a major role in the development of advanced technologies for hydrogen production.

6.1.1. *Synthesis of Catalysts*

It is well established that the solid catalysts for commercial applications need to be an active, selective, and stable material able to operate for periods between months and several years. The best synthesis method must be able to produce a catalytic material with appropriate textural properties (i.e., sufficiently high surface area and uniform pore distribution), suitable mechanical strength, and high attrition resistance. Within the context of the economic value of catalysis, it is estimated that the cost of the catalyst in a chemical process represents between 0.1% and 1% of the final product margin, whereas the total sales of catalysts are between \$15 and \$19 billion per year, which implicates that the total cost value generated by the catalysis industry can be on the order of trillions of dollars per year (Munnik *et al.*, 2015). These figures clearly show the massive impact that heterogeneous catalysis has had and will continue having over the energy and chemical industry economy in the future.

At lab scale, fine powder catalysts are usually used to prevent mass transfer limitations and to ensure good accessibility of the reactant(s) to the active sites. Unfortunately, small catalyst particles generate large pressure drop along the catalyst bed. Hence, large catalyst bodies of millimeter size are often employed in industry to overcome this limitation (Bartholomew and Farrauto, 2006). In Fig. 6.4, a trilobe-shaped catalyst with its cross section, mesoscopic level, and reactor dimensions' is shown to illustrate the multiple scales present in heterogeneous catalysis.

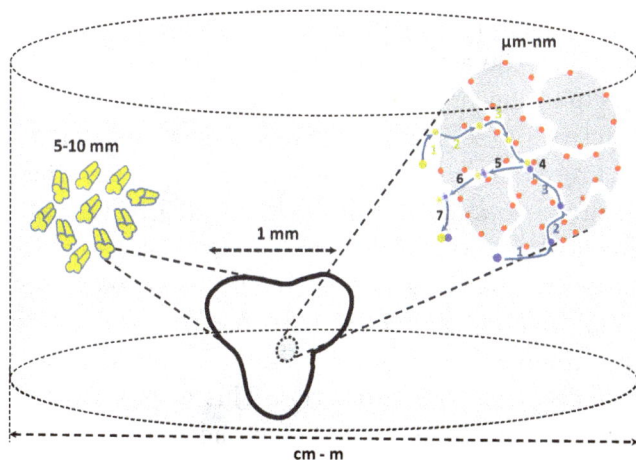

Figure 6.4. Length scales of reactor, metal catalyst structure, and sequence of physical and chemical reaction steps in heterogeneous catalysis. Adapted from González-Cortés and Imbert (2018).

The chemical and physical steps involved in the heterogeneous catalysis are also shown: (1) diffusion of reactants from the gas or liquid phase to the external surface of the porous catalyst particle, (2) intraparticle diffusion of the reactants (green and blue circles) through the catalyst pores to the internal active sites, (3) adsorption of reactants onto the metal sites, (4) chemical reaction on the catalyst surface, (5) desorption of products from the surface of the metal catalyst, (6) intraparticle diffusion of the products from the internal pores of the catalyst to the external surface of the catalyst particle, and (7) diffusion of the products from the external surface of the particle to the bulk of the fluid (Kapteijn *et al.*, 1993). The catalytically active sites (1–10 nm) inside the pores of support particles (μm) and the molecular transport occur at the mesoscopic length scale, whereas the chemical adsorption and reactions take place in the (sub)nanometer level. The pressure drops, mechanical strength, and attrition resistance are associated with the size and shape of the catalyst body; the scale is between centimeter (laboratory reactor) and meter (industrial reactor) (i.e., macroscopic length scale) (Chorkendorff and Niemantsverdriet, 2003).

The catalyst synthesis methodology groups a broad variety of methods in which the active phase is generated as a solid phase by either a precipitation or a decomposition reaction and methods in which the active phase is introduced and fixed onto a preexisting solid by a process intrinsically dependent on the surface of the support (Munnik *et al.*, 2015; Scharz *et al.*, 1995). The first category includes precipitation, coprecipitation, and sol-gel synthesis, among others, whereas the latter group is usually based on impregnation, adsorption, and deposition–precipitation methods (Regalbuto, 2007). The solution combustion synthesis (SCS) method can be classified within the abovementioned categories. This method was developed by Kingsley and Patil (1988) for the synthesis of metal oxides and consists of using a saturated aqueous solution of the desired metal salts (nitrates are generally preferred because of their oxidizing property and high solubility in water) and a suitable organic fuel as reducing agent (e.g., urea, glycine, carbohydrazide, and maleic hydrazide). The redox mixture is ignited and eventually combusted in a self-sustained and fast combustion reaction, resulting in a nanocrystalline oxide material (Bera, 2019; Carlos *et al.*, 2020; Patil *et al.*, 2008; Specchia *et al.*, 2010) or nanopowder of metal particles (Erri *et al.*, 2008; Khort *et al.*, 2017; Kumar *et al.*, 2011; Podbolotov *et al.*, 2017; Tappan *et al.*, 2006; Trusov *et al.*, 2016). In Chapter 1, we have already described that several factors such as metal precursors, fuels, ignition modes, solvents, and solution pH, among others, can have impact over the SCS of advanced catalysts and materials from a viewpoint of sustainability and greenness. It was proposed that a rational selection of synthesis parameters, such as (1) low decomposition and reduction temperature of abundant metal nitrates, (2) high solubility in water, (3) neutral acidity, (4) low tendency to contaminate the final catalyst or material, (5) minimization of noxious emissions during the combustion process (i.e., low gas emission factor), (6) utilization of waste-derived precursors and sustainable organic fuels, (7) ignition of redox mixture with microwave (or ultrasound), (8) high reducing valence of the fuel, particularly polycarboxilic acids and amino polycarboxilic acids, and (9) process intensification, among others, can enhance the greenness and sustainability of the SCS method in conjunction with the catalytic performance (i.e., activity, selectivity, and stability). This chapter, on the contrary, is focused on the applications of the SCS approach in various catalytic processes. The SCS of solid catalysts for photocatalysis, electro-catalysis, solid oxide fuel cell (SOFC) materials, and coatings has been already described in this book. Herein, emphasis is given on the development of technologies for producing hydrogen as alternative and clean fuel, using solution combustion (SC)-synthesized catalysts. We also include further applications of SC-synthesized catalysts for energy and environmental pollution control. It is taken into account the sustainability of the catalytic process and also the catalyst synthesis, reuse, and regenerability. All these aspects have strong influences over the impact the SCS will have over the sustainable development of advanced catalytic processes.

As mentioned above, there is a plethora of synthesis methods to prepare bulk and supported catalysts such as sol-gel, impregnation, deposition–precipitation, coprecipitation, and so on (de Jong, 2009; Delmon, 2007; Munnik *et al.*, 2015; Scharz *et al.*, 1995). Some of these approaches require the use of hazardous chemicals, expensive raw material, long thermal treatment, and/or several separation steps to get the final product. The SCS, on the other hand, is an alternative and relatively new catalyst preparation method able to produce advanced catalytic materials in a single-step or one-pot synthesis, whose catalytic applications are described below to illustrate the versatility of this synthesis approach and its potential advantages in terms of sustainability and greenness. Initially, the hydrogen production from the reforming of hydrocarbons has been comprehensively described; then, a brief overview of further applications of SC-synthesized catalysts for energy and environmental pollution control has been provided. The last part of the chapter includes an overall conclusion and a likely path forward of the SCS method.

6.2. Hydrogen Production from Hydrocarbon Fuels

Currently, commercial hydrogen gas is produced from the reforming of fossil fuels to mainly syngas ($H_2 + CO$) through three technologies: SR, PO, and ATR (Holladay *et al.*, 2009). It also produced carbon monoxide (CO), which is subsequently converted into CO_2 and hydrogen, through the water–gas shift reaction (WGSR), to increase the yield of hydrogen. Furthermore, CO is a poisonous gas for noble metal catalysts used in fuel cell, hence decreasing the efficiency of fuel cells. An interesting strategy to produce CO_x-free hydrogen is the nonoxidative catalytic conversion of either methane or liquid hydrocarbons to hydrogen and carbon nanofiber (Choudhary *et al.*, 2003; González-Cortés *et al.*, 2016; Jie *et al.*, 2019; Jie *et al.*, 2017; Muradov and Veziroglu, 2005).

A large variety of SC-synthesized catalysts such as noble metals, base metals — particularly nickel — and bimetallic formulations mainly supported on ceria and alumina, and also mixed-metal oxides as catalyst precursors has been investigated in the reforming of hydrocarbons. It is also an extended practice to produce the catalyst support by the SCS method and then deposit the metal precursors through wet impregnation. These studies have been carried out over different catalytic processes such as SR, dry reforming (DR), PO, ATR or oxy-steam reforming (OSR), tri-reforming (TR), WGSR, and even cracking or catalytic decomposition of hydrocarbons under nonoxidative atmosphere. The chemical reactions are shown in Equations (6.1)–(6.7) and their applications are described in the following sections:

Steam reforming:

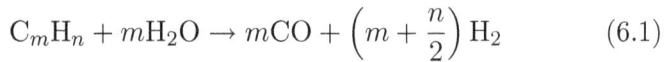

$$C_mH_n + mH_2O \rightarrow mCO + \left(m + \frac{n}{2}\right)H_2 \qquad (6.1)$$

Partial oxidation:

$$C_mH_n + \left(\frac{m}{2}\right)O_2 \rightarrow mCO + \left(\frac{n}{2}\right)H_2 \qquad (6.2)$$

Dry reforming:

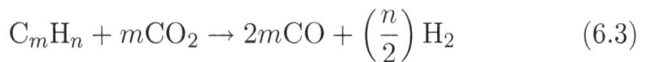

$$C_mH_n + mCO_2 \rightarrow 2mCO + \left(\frac{n}{2}\right)H_2 \qquad (6.3)$$

ATR or OSR:

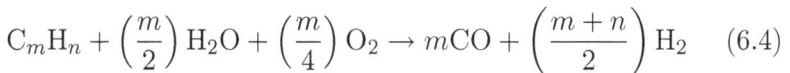

$$C_mH_n + \left(\frac{m}{2}\right)H_2O + \left(\frac{m}{4}\right)O_2 \rightarrow mCO + \left(\frac{m+n}{2}\right)H_2 \qquad (6.4)$$

Tri-reforming:

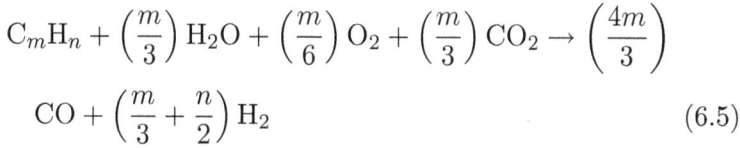

$$C_mH_n + \left(\frac{m}{3}\right)H_2O + \left(\frac{m}{6}\right)O_2 + \left(\frac{m}{3}\right)CO_2 \rightarrow \left(\frac{4m}{3}\right)$$

$$CO + \left(\frac{m}{3} + \frac{n}{2}\right)H_2 \qquad (6.5)$$

Water-gas shift reaction:

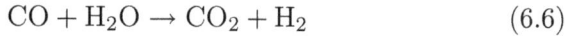

$$CO + H_2O \rightarrow CO_2 + H_2 \qquad (6.6)$$

Cracking or catalytic decomposition:

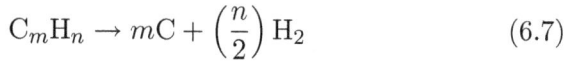

$$C_mH_n \rightarrow mC + \left(\frac{n}{2}\right)H_2 \qquad (6.7)$$

6.2.1. *Steam Reforming of Hydrocarbons*

The reforming of hydrocarbons, particularly SR (Eq. (6.1)), is actually the most widespread and least expensive industrial process for hydrogen production as a consequence of its relatively simple operation and low costs (Kalamaras and Efstathiou, 2013). The steam methane (natural gas) reforming process is carried out at high temperatures (700°C–1000°C) and pressure (3–25 bar) over a Ni catalyst to produce a gas stream mainly composed of hydrogen, CO, and CO_2. SR of hydrocarbons is an endothermic reaction that requires an external heat source. It operates at relatively lower temperature than PO (Eq. (6.2)) and ATR (Eq. (6.4)), and produces a hydrogen-enriched syngas with a high H_2/CO ratio (3:1). However, it also produces the highest carbon emissions compared to the PO and ATR.

The SCS method has been widely used to prepare catalysts for reforming of mainly methane but other hydrocarbons have also been examined. Alumina nanofiber (ANF)-supported NiO catalyst was synthesized by combustion synthesis through a series of stages as shown in Fig. 6.5.

The ANF impregnated with the aqueous redox mixture containing nickel (II) nitrate and glycine as fuel, glycine-to-metal ratio of ~0.82, was aged at room temperature for 30 min. Subsequently, it

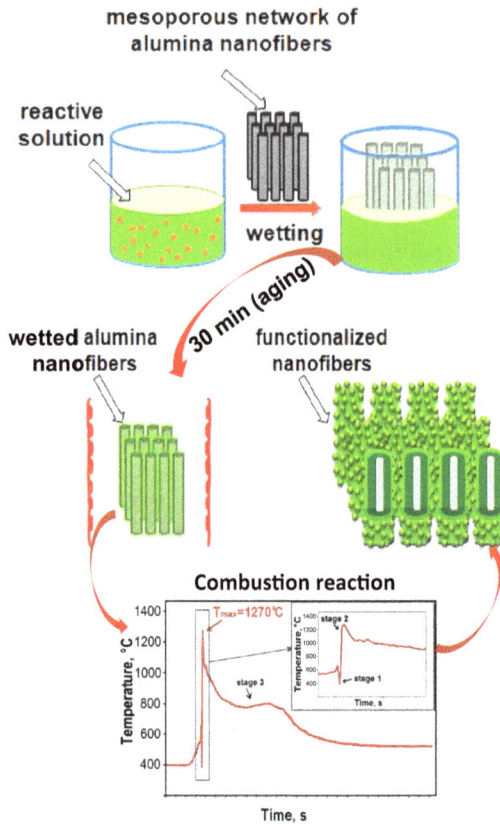

Figure 6.5.   Schematic representation of the SC-synthesized NiO/ANF catalyst precursor for steam reforming of methane (SRM) and CO_2 methanation. Adapted from Aghayan *et al.* (2017).

was placed into a muffle furnace preheated to ignition temperature of 400°C for 30 min to reach about 1270°C upon the combustion process. The advanced performance of Ni/ANF for steam reforming of methane (SRM) was associated with low density of the material, high gas permeability, high Ni content, and dispersion under high-temperature reaction conditions. The catalytic material was stable under operation conditions for at least 50 h of reaction, keeping a fibrous structure of the support unchanged and preventing sintering of Ni particles (Aghayan *et al.*, 2017). A similar synthesis procedure was used to make Ni-CeO_2/ANF catalyst, which exhibited a relative high activity for the reforming of propane at low steam-to-carbon ratio (i.e., 1) and a slow catalytic deactivation during 12 h of operation (Potemkin *et al.*, 2018).

The SCS has also been successfully applied to the preparation of catalyst supports used in the SRM. The supports are usually treated at high temperatures after the combustion process to burn off residual carbon and obtain the required crystalline structure. The SC-synthesized $Y_2Zr_2O_7$ pyrochlore-type structure using glycine as fuel was impregnated with an aqueous solution of nickel nitrate to produce 10% Ni/$Y_2Zr_2O_7$ catalyst after reducing under flowing hydrogen. This catalyst formulation exhibited superior activity, stability, and coke resistance than similar catalysts supported on $Y_2Zr_2O_7$ pyrochlore synthesized using hydrothermal and coprecipitation methods (Fang *et al.*, 2016). Another example is 4 wt% Ni/CaO–$Ca_{12}Al_{14}O_{33}$ catalyst, whose support (CaO–$Ca_{12}Al_{14}O_{33}$) with two different CaO/$Ca_{12}Al_{14}O_{33}$ weight ratios (i.e., 75/25 and 90/10) was prepared by microwave-assisted combustion method using urea as fuel. The two compositions of the CaO–$Ca_{12}Al_{14}O_{33}$ support (and sorbent) were impregnated with an aqueous solution of nickel nitrate to produce 4 wt% Ni/CaO–$Ca_{12}Al_{14}O_{33}$. The catalyst formulation with 75CaO/25$Ca_{12}Al_{14}O_{33}$ ratio was the most active and stable in CO_2 sorption-enhanced steam methane reforming. Its dual functionality as catalyst for steam methane reforming and simultaneous sorption of CO_2 over CaO shifts the equilibrium of the WGSR and hence enhancing the selectivity to about 100% hydrogen (Cesario *et al.*, 2015).

The Ir/CeO_2 catalysts with different loadings of Ir (0.1–1 wt%) were synthesized by SC using glycine as fuel. The SEM images for as-prepared CeO_2 and Ir/CeO_2 catalysts show the characteristic sponge-like microstructure for materials prepared by the SCS method (Fig. 6.6). This is because of the rapid evolution of gases during the exothermic combustion reaction. A step of calcination at 300°C followed by reduction at 500°C enhanced the catalytic performance of Ir/CeO_2 catalyst for the SRM in low-steam conditions. The 0.1

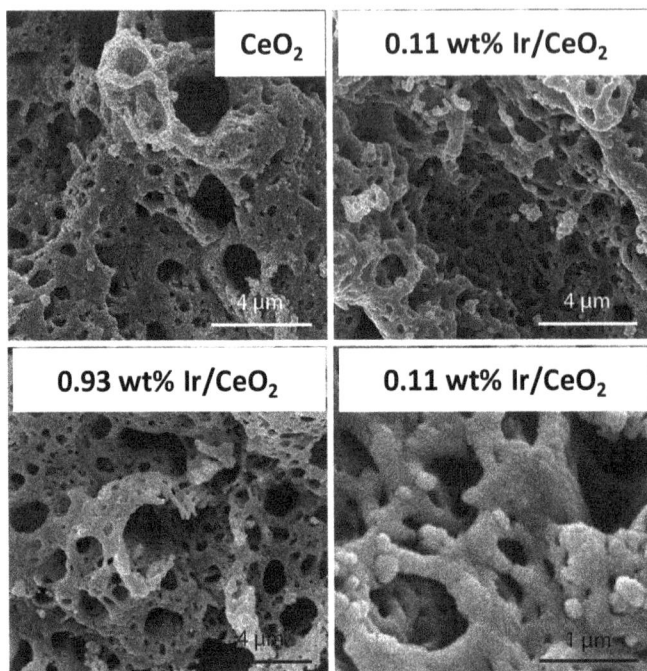

Figure 6.6.   SEM images for CeO_2 and Ir/CeO_2 catalysts prepared by the SCS. Adapted from Nguyen *et al.* (2014).

wt% Ir/CeO_2 formulation was successfully employed as catalyst for the production of hydrogen as a consequence of its higher Ir dispersion (smaller particle size, $\sim$2 nm) and stronger Ir–CeO_2 interaction (Nguyen *et al.*, 2014). The SC-synthesized Ir/CeO_2 catalysts were more active and stable than its counterpart synthesized by incipient wetness impregnation. Although the temporary collapse of the catalyst activity cannot be avoided when the catalyst was exposed to H_2S-containing stream, the Ir/CeO_2 almost fully recovers its initial activity on replacement of the sulfur-containing stream with the original sulfur-free methane/water mixture (Postole *et al.*, 2015). SC-synthesized Ni–MO_x (M = Ce, Fe, Al, Zr) catalysts using glycine as fuel were evaluated on the catalytic performance for SRM, after a calcination step at $800°C$ and a subsequent reduction process. It was found that Ni–AlO_x catalyst showed the best catalytic performance in terms of both methane conversion and hydrogen yield (Lima *et al.*, 2014). This finding could be because of the high surface area of the catalyst and a strong interaction between the nickel-active site and the aluminum oxide support. The utilization of Co–NiO–Al membrane prepared by the self-propagating high-temperature synthesis (SHS) method for methane reforming is also an interesting approach to produce hydrogen (Uvarov and Borovinskaya, 2013).

The SR of *n*-dodecane over CeO_2-supported noble metals has also been investigated. The CeO_2-supported 0.6 wt% Rh catalyst synthesized by combustion synthesis with urea as fuel was examined in the SR of *n*-dodecane. The high activity and stability for the SR of sulfur-free *n*-dodecane was attributed to the high metal dispersion and the strong metal-support interaction. Catalyst deactivation was observed in the presence of sulfur, mainly because of graphitic carbon deposition. Higher steam-to-carbon ratio enhanced the carbon gasification and the sulfur tolerance and hence the catalytic stability of the Rh/CeO_2 system (Vita *et al.*, 2017). The catalytic performance of SC-synthesized Pt/CeO_2 catalyst series with different Pt contents (between 0.6 and 2.3 wt%) was also evaluated in the SR of *n*-dodecane (Vita *et al.*, 2016). It was found that the optimal catalyst formulation (0.6 wt% Pt/CeO_2) showed high catalytic activity in terms of total *n*-dodecane conversion and high H_2 concentration (~73%). This catalyst showed stable performance during 50 h of reaction, with an activity similar to that of a commercial Rh catalyst. A catalyst bed configuration with temperature gradient (~480°C–800°C) decreased the coke formation especially at the inlet of the catalytic bed. Recently, this catalyst formulation was successfully evaluated in the SRM in cyclic operation (Amjad *et al.*, 2019). In summary, the SCS method has been applied to make not only SR catalysts but also catalyst supports, which are then impregnated with metal precursors to produce active SR catalyst. An additional thermal treatment over the combusted catalyst is currently a normal practice to burn off residual carbon material. Glycine as fuel and CeO_2 as support are often employed in the SC-synthesized SR catalysts.

6.2.2. *Dry Reforming of Methane*

The DR of hydrocarbons (Eq. (6.3)), especially methane (Eq. (6.8)), to CO and H_2 (synthesis gas or syngas) is an eco-friendly reaction able to convert simultaneously two GHGs (i.e., CO_2 and CH_4) to high-value products. Hydrogen can be used as alternative clean fuel and also as reactant for hydrogenation reactions and ammonia synthesis. Syngas, on the contrary, is the raw material for the production of liquid fuels through Fischer-Tröpsch (FT) synthesis and also methanol (Abdulrasheed *et al.*, 2019). The dry reforming of methane (DRM) to syngas (Eq. (6.8)) is an endothermic reaction that produces lower syngas ratio ($H_2/CO = 1$) than those for the SRM ($H_2/CO = 3$) and the PO of methane ($H_2/CO = 2$). This relatively lower reducing character of the reaction atmosphere and the high temperatures (650–850°C) favor several competitive side reactions such as methane cracking or decomposition (Eq. (6.9)), reverse WGSR (Eq. (6.10)), and the reverse reaction of CO dispropor-tionation (Boudouard reaction) (Eq. (6.11)). The side reactions are taken place when different CO_2 and CH_4 conversions are obtained, the H_2/CO ratio does not match with the ideal stoichiometric proportion, water is detected in the outlet stream, and carbon is deposited on catalyst surface:

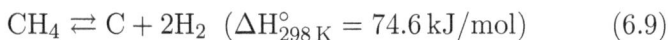

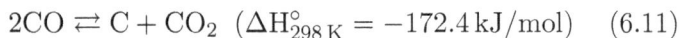

$$CH_4 + CO_2 \rightleftarrows 2CO + 2H_2 \quad (\Delta H^{\circ}_{298\,K} = 247.0\,kJ/mol) \qquad (6.8)$$

$$CH_4 \rightleftarrows C + 2H_2 \quad (\Delta H^{\circ}_{298\,K} = 74.6\,kJ/mol) \qquad (6.9)$$

$$CO_2 + 2H_2 \rightleftarrows CO + 2H_2O \quad (\Delta H^{\circ}_{298\,K} = 41.1\,kJ/mol) \qquad (6.10)$$

$$2CO \rightleftarrows C + CO_2 \quad (\Delta H^{\circ}_{298\,K} = -172.4\,kJ/mol) \qquad (6.11)$$

Ali and coworkers compared the physicochemical properties, chemical structure, and catalytic performance for the DRM of 5 wt% Ni/Al_2O_3 catalysts prepared by the SCS method and wet impregnation (Ali *et al.*, 2019). Figures 6.7a and 6.7b clearly exhibits the superior CH_4 and CO_2 conversions for SC-synthesized Ni catalyst compared to wet impregnation catalyst. The combustion-prepared catalyst also showed higher yields of hydrogen and CO and markedly superior catalytic stability than impregnation-prepared catalyst. This finding was attributed to the smaller Ni particle

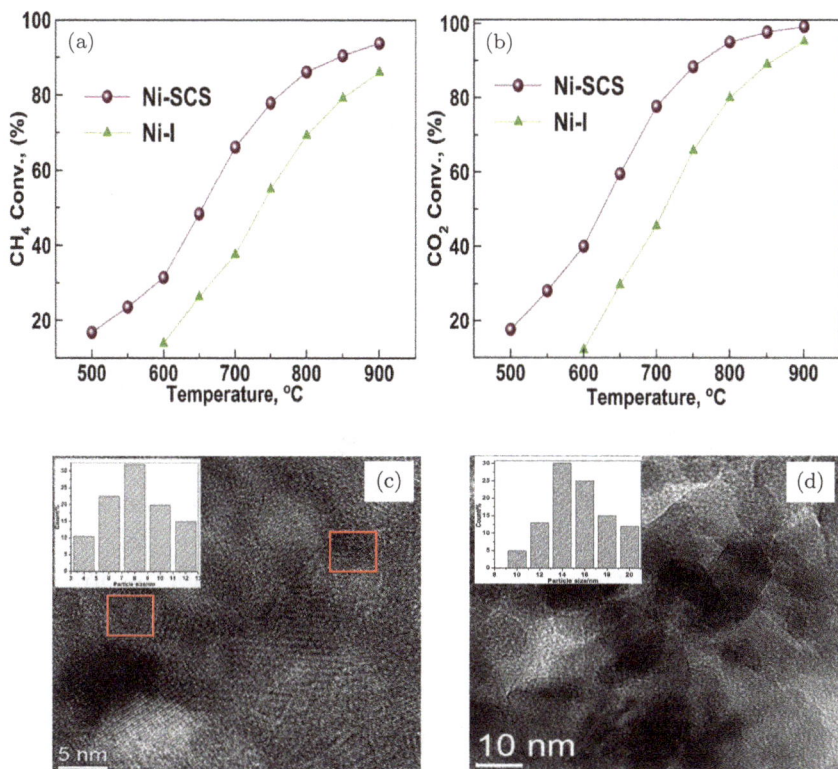

Figure 6.7. DRM over 5 wt% Ni/Al₂O₃ catalyst. (a) Methane conversion, (b) CO₂ conversion, (c) HRTEM images and metal particle distributions for calcined NiO/Al₂O₃ catalyst precursor prepared by the SCS method, and (d) wet impregnation. Reaction conducted over 0.5 g of catalyst, CH4:CO₂:Ar volume ratio of 1:1:1 and 120 mL/min. Adapted from Ali *et al.* (2019).

size (Figs. 6.7c and 6.7d), better dispersion, stronger metal support interaction (SMSI) effects, and the presence of reducible NiAl₂O₄ nano-crystallites in the SC-synthesized Ni catalyst.

The Pr-promoted Ni−Mg−Al catalysts prepared by microwave-assisted combustion of metal (Mg, Al, Ni, and Pr) nitrates and glycine as fuel were also evaluated in the dry reforming of methane reaction. It was found that Pr-Ni−Mg−Al catalysts were active for the hydrogen production reaction. The presence of Pr improved the basic characteristics of the material allowing more efficient carbon gasification. A loading of 6 wt% Pr improved the catalytic stability and decreased the formation of carbon on the DRM

(Ojeda-Nino *et al.*, 2019). The SCS method was also used to prepare $Ni/Nd_4Ga_2O_9$ catalysts containing 5, 10, and 15 wt% of Ni. The best catalytic activity for the DRM was obtained at 800°C for the catalyst with a Ni content of 10 wt% and reduced at 700°C. This spent catalyst showed Ni particles between 10 and 20 nm indicating that little sintering occurred under reaction conditions. Moreover, no carbonaceous deposits were observed after 12 h on steam (Garcia *et al.*, 2008).

The SHS of Ni_3Al and subsequent deposition of Pt (or Ru) and the increase of the metal content strongly enhanced the CH_4 and CO_2 conversions. The most active Pt/Ni_3Al catalyst did not undergo deactivation because of coking. This catalyst was active and stable during 120 h of time on stream (Arkatova *et al.*, 2018). Nickel aluminide coatings on substrates also showed high activity and syngas selectivity for the DRM (Xanthopoulou *et al.*, 2017). This reaction was also carried out over various Ni–Al membrane prepared by SHS method and modified by nanosized metal–metal oxides uniformly distributed inside membrane pores (Tsodikov *et al.*, 2011). SHS method was also used to prepare monolith-based catalysts able to convert selectively methane and natural gas to syngas (Maksimov *et al.*, 2013).

The SCS of mixed-metal oxides as catalyst precursors for the reaction of DRM is a well-established strategy to develop efficient catalysts. Perovskite-type $La(Co_xNi_{1-x})_{0.5}Fe_{0.5}O_3$-mixed oxides with different Co and Ni contents were synthesized by SC using citric acid monohydrate (CA) as fuel and metal (La, Co, Ni, and Fe) nitrates with a CA/metal ratio of 1.25 and pH of 8. The aqueous redox mixtures were ignited to obtain spongy-like materials, which were then calcined at 700°C for 5 h to obtain the powder catalyst precursors. Bimetallic Ni–Co catalysts supported on La_2O_3–$LaFeO_3$ were prepared by reducing the $La(Co_xNi_{1-x})_{0.5}Fe_{0.5}O_3$-mixed metal oxides (Wang *et al.*, 2019). The catalytic activity and coking resistance were enhanced by substituting a proper amount of Co as a consequence of the synergistic effect between Ni and Co. Catalysts reduced from orthorhombic perovskite precursors with $x = 0.10$ and 0.30 showed higher structural stability, maximal catalytic activity,

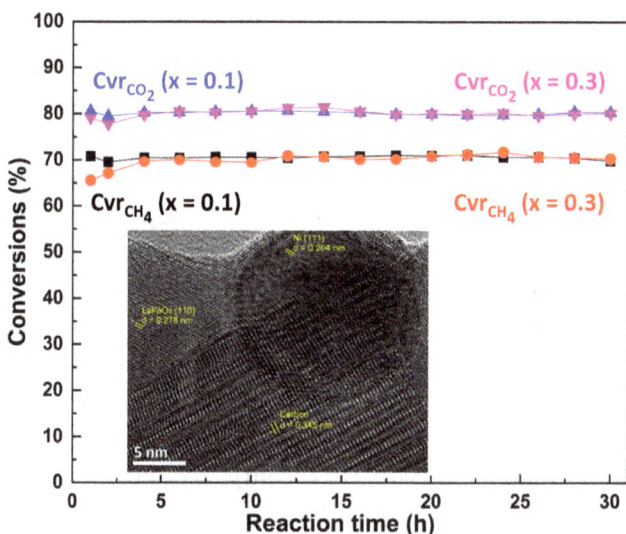

Figure 6.8. Dependence of the conversions of CH_4 and CO_2 with time on stream for DRM over $La(Co_{0.1}Ni_{0.9})_{0.5}Fe_{0.5}O_3$ ($x = 0.1$) and $La(Co_{0.3}Ni_{0.7})_{0.5}Fe_{0.5}O_3$ ($x = 0.3$). Inset corresponds to HRTEM image of the spent $La(Co_{0.1}Ni_{0.9})_{0.5}Fe_{0.5}O_3$ catalyst. Adapted from Wang *et al.* (2019).

and coking resistance as shown in Fig. 6.8. It was also found that the crystalline structure of the catalyst precursor plays an important role in the catalyst performance.

The Ni–Cu–La_2O_3 catalyst obtained after reducing the SC-synthesized $La_2(Ni_{0.8}Cu_{0.2})O_4$ catalyst precursor showed an optimal performance for the DRM compared to $La_2(Ni_{1-x}Cu_x)O_4$ ($x = 0.0, 0.1, 0.3$, and 0.4). The enhanced carbon resistance of this catalyst formulation compared to an equivalent catalyst prepared by wet impregnation was associated with the smaller metallic particles of the reduced perovskite and the cage-like structure of surface-segregated Cu (Wang *et al.*, 2017). SC-synthesized $LaNiNb$ mixed-metal oxides with small contents of Nb (Alvarez *et al.*, 2011) and $LaSrNiAl$ perovskite-type structures (Rodriguez-Sulbaran *et al.*, 2018) also showed remarkable catalytic activity and stability during the DRM.

The catalytic formulation for the DRM of SC-synthesized $La_xBa_{1-x}Ni_yCu_{1-y}O_3$ perovskite-type oxides was optimized using three-layer artificial neural network model based on catalyst preparation parameters and calcination temperature coupled with genetic algorithm (Abbasi *et al.*, 2017). SC-synthesized $Ce_{0.70}La_{0.20}Ni_{0.10}O_{2-\delta}$ type-mixed-metal oxide resulted to be a good catalyst precursor for the DRM reaction. Strong Ni-support interaction was enhanced by rising the prereduction temperature of the mixed-metal oxide (from 450°C to 900°C). A good balance between Ni reduction degree and metal-support interactions after reduction at 750°C is associated with the high resistance to carbon deposition (Pino *et al.*, 2017). According to the above-reviewed literature, a variety of Ni-containing mixed-metal oxides, particularly La-based perovskites, as catalyst precursors have been synthesized by SC and examined in the DRM reaction. In general, the catalysts have exhibited high activity, selectivity to syngas, and catalytic stability because of nanosized Ni particles and strong Ni-support interactions. The SCS approach has also been used to make active Ni catalysts supported on alumina (or periclase-like structure). The nickel aluminide (Ni_xAl_y) catalysts prepared by SHS method and applied in the DRM reaction have also showed advanced catalytic performances.

6.2.3. *Partial Oxidation of Methane*

The gasification of a hydrocarbon in the presence of oxygen (PO of hydrocarbons) is shown in Eq. (6.2). Other reactions can also occur as shown in the PO of methane (PMO) (Eq. (6.12)) through the total combustion reaction (Eq. (6.13)) and possibly SR (Eq. (6.14)). The PMO is a noncatalytic process that is carried out at temperatures in the 1300°C–1500°C range and pressures in the 3–8 MPa range (Kalamaras and Efstathiou, 2013):

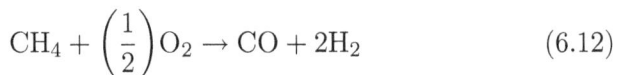

$$CH_4 + \left(\frac{1}{2}\right)O_2 \rightarrow CO + 2H_2 \qquad (6.12)$$

$$CH_4 + 2O_2 \rightarrow CO_2 + 2H_2O \qquad (6.13)$$

$$CH_4 + H_2O \rightarrow CO + 3H_2 \qquad (6.14)$$

The catalytic partial oxidation (CPO) of methane over noble metal (Pt, Rh, Ir, and Pd) and non-noble metal (Ni and Co) catalysts (Enger *et al.*, 2008; York *et al.*, 2003) takes place under less severe

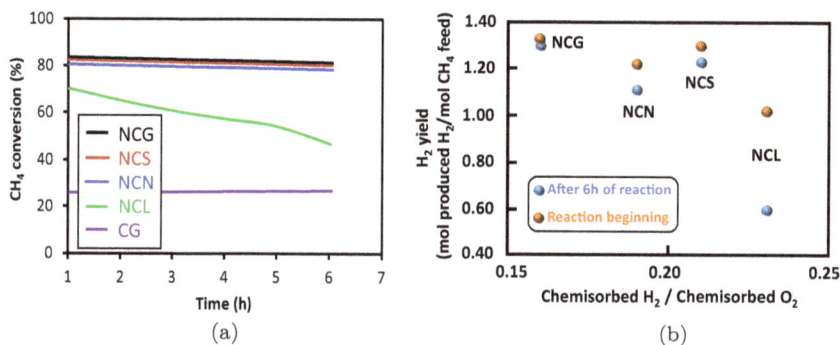

Figure 6.9. Dependence of the CH_4 conversions with time on stream for the PO of methane over $Ni_{0.1}(Ce_{0.9}Ln_{0.1}O_{1.95})_{0.9}$ catalysts, where (a) Ln = Gd (NCG), La (NCL), Nd (NCN), Sm (NCS), and $Ce_{0.9}Gd_{0.1}O_{1.95}$, CG. (b) H_2 yield versus chemisorbed H_2 (catalysts)/chemisorbed O_2 (supports) (mol/mol). Adapted from Alvarez-Galvan *et al.* (2018).

operation condition than the POM reaction, whose contact time (between 1 and 40 ms) is several orders of magnitude shorter than SR. These advantages make CPOM a promising technology for the upcoming gas economy as it allows compact, economic, and high-energy-efficient reactors, which promote a decentralized natural gas utilization and hydrogen production (Al-Sayari, 2013).

SC-synthesized nickel nanoparticles supported on highly crystalline Ce–lanthanide–mixed oxides are active catalysts for hydrogen production from the PO of methane (Alvarez-Galvan *et al.*, 2018). The best catalytic performance corresponded to the $Ni/Ce_{0.9}Gd_{0.1}O_{1.95}$ catalyst (Fig. 6.9a), in which low CO_2 selectivity and no carbon formation were observed. It was found that the ratio between Ni dispersion and the oxygen chemisorption capacity of the doped-ceria support plays a crucial role in the catalyst reactivity for POM to syngas (Fig. 6.9b) in agreement with a previous report of Gd-doped ceria ($Ce_{1-x}Gd_xO_{2-\delta}$) catalyst support prepared by the SCS method (Hennings and Reimert, 2007).

The CPOM reaction has also been studied over $Ce_{0.9}Zr_{0.1}O_2$-supported Ni and Cu–Ni catalysts by in-situ X-ray absorption near-edge structure (XANES) technique. It was found that the $Ni/Ce_{0.9}Zr_{0.1}O_2$ catalyst showed a continuous reduction of Ce^{4+} cation in CH_4–O_2 feed flow, carbon formation, and Ni reoxidation took place under catalyst operation condition. In contrast, $Cu–Ni/Ce_{0.9}Zr_{0.1}O_2$ catalyst displayed a slight reoxidation of cerium oxide and no reoxidation of Ni alongside the suppression of carbon formation, indicating that the presence of copper enhances the catalytic stability because of an excellent resistance to carbon formation (Toscani *et al.*, 2018). Pino and coworkers (Pino *et al.*, 2003) carried out a comparative study of 1 wt% Pt/CeO_2 catalysts prepared by different methods and evaluated in the CPO of methane to syngas. The 1 wt% Pt/CeO_2 catalyst prepared by the SCS method and impregnation showed higher methane conversion and catalytic stability during 100 h of reaction compared to a similar catalyst prepared by coprecipitation. The strong metal–ceria interaction and the formation of solid solution can contribute to the high activity and stability of the Pt/CeO_2 catalysts. It has also been explored the utilization of perovskite-type $La_{1-x}Sr_xFeO_3$ synthesized by SC method for hydrogen generation. It was found that $La_{0.7}Sr_{0.3}FeO_3$ is an effective redox catalyst for methane PO and thermochemical water splitting in a cyclic redox scheme conducted at 850°C during 10 successive redox tests (He *et al.*, 2019). In summary, the SC-synthesized Ni catalysts for the CPOM reactions have been mainly supported on CeO_2 or modified CeO_2. Transition metals such as Fe, Pt, and even Cu-promoted Ni have also been used as catalysts. A mixed-metal oxide is usually used as catalyst precursor that releases the metal catalyst on the reduction process to enhance the catalyst performance in terms of conversion, hydrogen selectivity, and catalytic stability.

6.2.4. *Dual Reforming, Tri-Reforming, and Catalytic Cracking of Methane*

The dual reforming of hydrocarbons can be a combination of SR and PO of methane called ATR (Eq. (6.4)) or even combined DR and PO of methane usually called dry-oxy-reforming (DOR) of methane (Eq. (6.15)):

$$2CH_4 + CO_2 + H_2O \rightarrow 3CO + 5H_2 \qquad (6.15)$$

TR of hydrocarbons (Eq. (6.5)) is the combination of the SR, PO, and DR reactions. On the contrary, the cracking of hydrocarbons (Eq. (6.7)), particularly methane (Eq. (6.9)), generates mainly hydrogen and solid carbon as main products. The utilization of a solid catalyst decreases markedly the reaction temperature and tunes the selectivity to hydrogen. These combined reactions have been widely examined over SC-synthesized catalysts as discussed in the following paragraphs.

The CeO_2-supported $\sim$1 wt% Pt solid prepared by SC method is an active and stable catalyst for the ATR of propane. This solid catalyst was tested during several hours in a wide range of gas hourly space velocity. The catalyst characterizations revealed the presence of platinum as metal and ionically substituted state in the ceria support with the absence of carbon deposition in the spent catalyst (Pino *et al.*, 2006). The coating of SC-synthesized Pt/CeO_2 catalyst over a monolithic structure also produced high catalytic performance in terms of fuel conversion, hydrogen production, and low by-products formation (Vita *et al.*, 2010). This catalytic system (i.e., Pt/CeO_2) synthesized by SC was also successfully utilized in a hydrogen-generation prototype to catalyze the ATR of methane (Cipiti *et al.*, 2013). The utilization of liquid fuels as feedstock for hydrogen generation was studied by Varma *et al.* (2006). They found that the SC-synthesized $La_{1-y}Ce_yFe_{1-x}Ni_xO_3$ catalyst precursors showed phase segregation with Ce exhibiting limited solubility and Ni separating primarily during the initial reduction stage. The Fe–Ni–CeO_2–La_2O_3 catalysts showed excellent stability while maintaining above 90% conversion and near-equilibrium hydrogen production for the ATR of liquid (JP-8) fuel, even at high GHSV values (130,000 h^{-1}).

In a recent study of the DOR of methane, the sol-gel plasma-prepared Co-promoted $NiAl_2O_4$ catalyst showed higher CH_4, CO_2 conversions, and stability than the sol-gel, and even SC-prepared catalysts as a consequence of its small particle size, high surface area, and reducibility (Sajjadia and Haghighi, 2019). The TR of methane over lanthana-promoted Ni–CeO_2 catalysts prepared by the SCS method revealed that a strong interaction between nickel, lanthana, and surface oxygen vacancies of ceria can induce the formation of Ce^{3+} sites and hence enhancing the Ni dispersion and the catalytic activity. Furthermore, the lanthana addition improved the interactions of CO_2 with the catalyst surface and hence the catalytic activity (Pino *et al.*, 2011).

The SC-synthesized materials for methane cracking reaction are mainly Ni-based catalysts. In this context, Kuvshinov *et al.* (2019) examined the influence of the heating rate on the SCS of 97% NiO-3% Al_2O_3 catalysts. The redox mixtures containing hexamethylenetetramine (HMT) as fuel, Ni, and Al nitrates were heated from room temperature to 450°C at different heating rates (i.e., 1, 10, and 15 K/min) to obtain three catalysts. The Ni-based catalysts showed a mixture of NiO and Ni crystalline phases, whose abundance depends on the heating rate. The NiO–Ni–Al_2O_3 catalyst synthesized at higher heating rate and the enriched Ni nanoparticles achieved the maximum combustion temperature, the maximum specific yield of hydrogen from methane cracking, and the maximum catalytic stability as shown in Fig. 6.10.

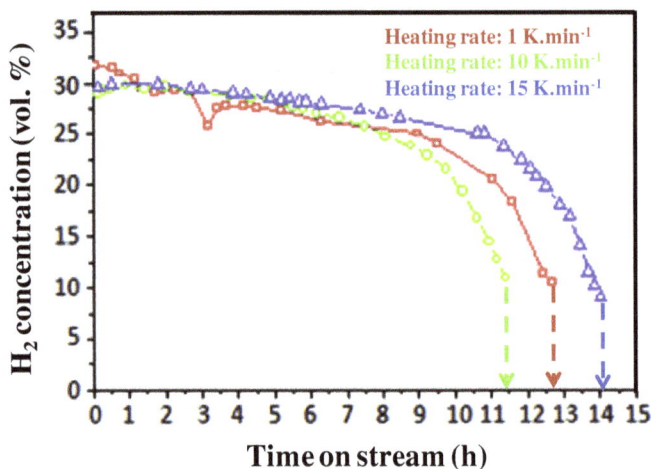

Figure 6.10. Hydrogen concentration versus time on stream for the methane cracking reaction over SC-synthesized Ni catalysts treated at different heating rates. Reactions are conducted at atmospheric pressure, 550°C, and methane-specific flow rate of 100 L/h.g_{cat}. Adapted from Kuvshinov *et al.* (2019).

SC-synthesized SiO$_2$-supported Ni–La$_2$O$_3$ catalysts were also effective to produce hydrogen from methane cracking when SiO$_2$ support with high surface area was used to produce well-dispersed Ni nanoparticles (Tajuddin *et al.*, 2019). The Pt-promoted Ni–CeO$_2$ catalysts prepared by the impregnation of SC-synthesized CeO$_2$ with Ni and Pt precursors were evaluated in the methane cracking reaction (Pudukudy *et al.*, 2018). The increase of Pt loading and the reaction temperature enhanced the hydrogen yield and catalyst stability as a consequence of the well-dispersed Ni and Pt over CeO$_2$ and a proper metal-support interaction. Furthermore, the textural properties of the residual multi-walled carbon nanotube changed considerably with the catalyst formulation. The DR and TR of methane alongside the methane cracking reaction have been investigated through SC-synthesized Ni catalysts supported mainly on CeO$_2$, but other supports such as Al$_2$O$_3$ and SiO$_2$ have also been used. Promoters such as La$_2$O$_3$, Pt, and Fe have been used to enhance the generation of hydrogen and the catalyst stability.

6.3. Further Applications of SC-Synthesized Catalysts

The SCS method has been widely used in the preparation of a large variety of solid catalysts and even catalyst carriers or supports for numerous catalytic reactions as listed in Table 6.1. For instance, SC-synthesized catalysts have been applied for hydrogen production through reforming not only of fossil-derived feedstocks as discussed but also of sustainable raw materials such as ethanol (Cross *et al.*, 2019; Luiz *et al.*, 2016), biogas (Vita *et al.*, 2018), and even methanol (Ajamein *et al.*, 2017; Khani *et al.*, 2018). Hydrogen production was also examined over SC-synthesized CeO$_2$ carrier for Rh catalyst in the WGSR (Eq. (6.6)) (Galletti *et al.*, 2011). In general, noble metal ionic catalysts (i.e., Ce$_{1-x}$M$_x$O$_{2-\delta}$) synthesized by the SCS method have showed very good activity for the WGSR (Bera, 2019). Another example of sustainable application of the SCS method is the biodiesel production through the transesterification reaction of vegetable oils over SC-synthesized catalysts (Lamba *et al.*, 2019; Nayebzadeh *et al.*, 2019; Yousefi *et al.*, 2019).

The utilization of structured catalysts for energy applications and pollution control is an additional and active research area, where the SCS method has favorably impacted because of its intrinsic energy saving and good performances of the SC-synthesized catalysts in different applications (Specchia *et al.*, 2017). The cordierite monoliths-coated $Ni/Ce_{0.8}Gd_{0.2}O_2$ catalyst prepared by the SCS method is an effective approach to produce methane from CO_2 hydrogenation reaction (Vita *et al.*, 2018) as given in Table 6.1. A similar methodology was used by Wang *et al.* (2018) to coat MnO_x–CeO_2–Al_2O_3 on cordierite honeycomb (monolith) for the selective catalytic reduction of NO_x with NH_3.

The hydrodesulfurization (HDS) of thiophene over wet impregnation SC-synthesized CoMo, NiMo, and NiW catalysts was previously investigated (González-Cortés *et al.*, 2014; González-Cortés *et al.*, 2014; González-Cortés *et al.*, 2006; González-Cortés *et al.*, 2004; González-Cortés *et al.*, 2006; González-Cortés *et al.*, 2005). In general, it was found that the thiophene conversion and selectivity to C_4 molecules are enhanced when the HDS catalysts are prepared by the SCS approach. Recently, it was reported that SC-synthesized $BaFe_{18}O_{27}$ catalyst is also active for the HDS of liquid fuels (Mandizadeh *et al.*, 2017); see Table 6.1.

The partial and total oxidation reactions of hydrocarbons alongside the soot combustion have been also studied over SC-synthesized solid catalysts. For instance, the selective oxidation of *n*-butane to maleic anhydride over $(V_{1-x}W_x)OPO_4$ catalysts has been successfully carried out (Schulz *et al.*, 2019). The total combustion of butane was also examined over MnO_x, MnO_x–La_2O_3, and MnO_x–CeO_x monolithic catalysts synthesized by SC (Shikina *et al.*, 2019). A greener SCS approach is based on the utilization of bio-extract, mainly phenolic-type acids (i.e., fuel), whereas the mineral nutrients (Na^+, K^+, Ca^{2+}, and Mg^{2+}) together with nitrate as combustion-aid agents assist the synthesis of active $LaCoO_3$ perovskite-type catalyst for the benzene oxidation reaction (Wang *et al.*, 2019); see Table 6.1. Another interesting SCS approach is the utilization of silica as hard template on the combustion synthesis of NiO nanocatalyst. Silica confines the combustion reaction in small nano-domains and

Table 6.1. Additional sustainable and unsustainable applications of SC-synthesized catalysts.

Application	Catalyst (fuel)	Major findings	Reference
Biohydrogen production	Ni (and/or Cu)/SiO$_2$ or CeO$_x$ (glycine)	The SCS-assisted impregnation is an effective method to prepare highly active, selective, and stable SiO$_2$-supported Ni and Ni–Cu catalysts to produce hydrogen from the catalytic conversion of ethanol	Cross *et al.* (2019)
WGSR	Rh/CeO$_2$ (not reported)	The Rh/CeO$_2$ catalysts using SC-synthesized CeO$_2$ showed significantly higher catalytic activity than a high-temperature-synthesized catalyst formulation	Galletti *et al.* (2011)
Biodiesel production	MgO/ MgAl$_2$O$_4$ (glycine)	The transesterification reaction of sunflower oil with methanol over SC-synthesized MgO/MgAl$_2$O$_4$ catalyst showed high reusability and conversion, which decreased less than 10% after five cycles of reactions	Yousefi *et al.* (2019)
CO$_2$ methanation	Ni/Ce$_{0.8}$Gd$_{0.2}$O$_2$ (urea)	The uniform, thin, and high-resistance layers of 50 wt% Ni/Ce$_{0.8}$Gd$_{0.2}$O$_2$ deposited on the cordierite monoliths by the SCS method showed high methane productivity and catalytic stability during 200 h of time on stream	Vita *et al.* (2018)

(*Continued*)

Table 6.1.   *(Continued)*

Application	Catalyst (fuel)	Major findings	Reference
Selective catalytic reduction of NO_x with NH_3	MnOx– CeO_2– Al_2O_3 (urea)	The SC-synthesized powder catalyst (i.e., MnO_x–CeO_2–Al_2O_3) and the cordierite honeycomb-coated MnO_x–CeO_2–Al_2O_3 catalyst were active for the selective catalytic reduction of NO_x with ammonia	Wang *et al.* (2018)
HDS of liquid fuels	$BaFe_{18}O_{27}$ (glucose, fructose, and maltose)	The SC-synthesized $BaFe_{18}O_{27}$ nanoparticles are efficient catalysts for HDS of liquid fuels. The ferrite-type catalyst was successfully regenerated after calcination at $500°C$ in air	Mandizadeh *et al.* (2017)
Benzene oxidation	$LaCoO_3$ (bio-extract)	The SC-biosynthesized $LaCoO_3$ catalyst exhibited an excellent catalytic performance for the benzene oxidation, in which a stable conversion above 90% at $285°C$ in a continuous run for 80 h was achieved	Wang *et al.* (2018)
Soot oxidation	NiO (glycine)	High catalytic activity for soot oxidation was achieved at $394°C$ using SC-synthesized NiO with three-dimensional open interconnected meso/macroporous structure and high concentration of oxygen vacancies	Voskanyan and Chan (2018)

favors an ultrafast cooling process that inhibits an extended crystal growth and hence facilitating the catalytic performance for the soot oxidation reaction (Voskanyan and Chan, 2018).

6.4. Conclusion and Path Forward

The green SCS methodology has had an impressive impact in catalyst preparation science over the last two decades. The applications in catalysis have been extended very rapidly owing to its facility to produce nanostructured mixed-metal oxide catalysts and even metal nanocatalysts through a one-pot synthesis approach. The solid catalysts have shown not only higher catalyst activity–selectivity but also higher catalyst stability and reusability.

The current contribution of SC approach in the production of hydrogen and environmental pollution control involves a variety of catalysts such as noble metals, nickel, and bimetallic formulations mainly supported on ceria and alumina. These catalysts are mainly obtained from SC-synthesized mixed-metal oxides as catalyst precursors, which after reduction produce nanosized metal particles and present strong metal-support interactions that enhance the activity, hydrogen selectivity, and catalytic stability. It has become an extended practice to make the catalyst support by SC method and then depositing the metal precursor(s) by impregnation. The vast majority of the catalyst formulations has been examined in the reforming of hydrocarbons, WGSR, and even cracking (or catalytic decomposition) of methane under nonoxidative atmosphere. SC-synthesized nanopowder catalysts or structured catalysts have also showed advantages for bio-hydrogen production, methane oxidation, soot combustion, HDS, and NO_x abatement among many others applications.

In the context of this book, we have comprehensively examined the advantages and challenges of the SCS method from the viewpoint of greenness and sustainable development of innovative catalytic processes. Several synthesis factors such as metal precursors, fuels, ignition modes, solvents, hard templates, and solution pH, among others, have shown strong influences over the greenness of the SCS of advanced catalysts and materials. It has been also surveyed the principles, the applications, and the most significant progress in the microwave-assisted SCS of nanostructured catalysts, which exhibited high surface area and porosity while compared to other conventional methods. The catalytic activities of the SC-synthesized materials offered enhanced catalytic performance.

A wide range of photocatalytic materials such as metal oxides, doped-metal oxides, and supported metal oxides synthesized by SC route were extensively investigated for photocatalytic applications. Primary emphasis has been given on the synthesis of nanostructured photocatalysts to control different properties such as the bulk structure, preferential formation of one polymorph over the other, surface morphology, crystallinity, particle size, and so on. Eventually, these properties enhance the photocatalytic efficiency of the materials under visible light exposure.

A growing number of catalytic materials synthesized by SC method are very attractive for electrochemistry applications, owing to the high specific surface area, the nanocrystalline nature of the obtained products, and its capability to synthesize metastable compounds. They have been widely used in oxygen reduction reaction (ORR), oxygen evolution reaction (OER), hydrogen evolution reaction (HER), photoelectrochemical (PEC), electrocatalytic ethanol oxidation, and supercapacitors.

Various syntheses routes, such as solid-state method, coprecipitation, sol-gel, SC method, and so on, have been used for the synthesis of SOFC materials. Among these methods, the SC process is the most popular method that is employed for the synthesis of an array of oxide materials for SOFC application because of its simplicity and versatility. SC method has been used for the synthesis of cathode, anode, electrolyte, and interconnect materials using different fuels, the mixture of fuels, and varying oxidizer-to-fuel ratio.

Envisioning a path forward for SC method applied in catalysis needs to be linked to the development of advanced technologies that meet the standards of greenness and sustainability to protect the human health, the environment, and at the same time fuel our society. Within this spirit, it is paramount to utilize sustainable or carbon neutral fuels instead of fossil-derived fuels to minimize (or eliminate) the environmental impact of anthropogenic emissions on the SCS of nanostructured catalysts. It is also worthwhile to minimize the energy consumption to achieve the ignition temperature of the redox mixture by utilizing microwave dielectric heating (or high-frequency heating), ultrasound, or a sustainable source of energy. An important challenge of the SC method, particularly during the scale-up stage, is to control the large generation of heat during the combustion process by a proper selection of the fuel, fuel-to-oxidizer ratio, and initiation mode of the combustion process.

In terms of the catalytic reaction, it is also highly decidable that it meets the criterions of greenness. Of course, the best possible scenario would be the utilization of a holistic approach in which the SC-synthesized catalyst alongside the catalytic process is integrally considered, designed, and implemented within the context of the 12 principles of Green Chemistry.

Acknowledgments

The author thanks the University of Oxford for generous funding and continual support of this multidisciplinary research.

References

Abbasi, M., Niaei, A., Salari, D., Hosseini, S. A., Abedini, F., & Marmarshahi, S. (2017). Modeling and optimization of synthesis parameters in nanostructure La1-xBaxNi1-yCuyO3 catalysts used in the reforming of methane with CO2. *J. Taiwan Inst. Chem. E.,* *74*, 187–195.

Abdulrasheed, A., Jalil, A. A., Gambo, Y., Ibrahim, M., Hambali, H. U., & Hamid, M. Y. S. (2019). A review on catalyst development for dry reforming of methane to syngas: Recent advances. *Renew. Sust. Energ. Rev.,* *108*, 175–193.

Aghayan, M., Potemkin, D. I., Rubio-Marcos, F., Uskov, S. I., Snytnikov, P. V., & Hussainova, I. (2017). Template-assisted wet-combustion synthesis of fibrous nickel-based catalyst for carbon dioxide methanation and methane steam reforming. *ACS Appl. Mater. Interfaces,* *9*, 43553–43562.

Ajamein, H., Haghighi, M., Alaei, S., & Minaei, S. (2017). Ammonium nitrate-enhanced microwave solution combustion fabrication of CuO/ZnO/Al2O3 nanocatalyst for fuel cell grade hydrogen supply. *Micropor. Mesopor. Mat.,* *245*, 82–93.

Al-Sayari, S. A. (2013). Recent developments in the partial oxidation of methane to syngas. *Open Catal. J,* *6*, 17–28.

Ali, S., Khader, M. M., Almarri, M. J., & Abdelmoneim, A. G. (2019). Ni-based nano-catalysts for the dry reforming of methane. *Catal. Today*, https://doi.org/10.1016/j.cattod.2019.1004.1066.

Alvarez-Galvan, C., Falcon, H., Cascos, V., Troncoso, L., Perez-Ferreras, S., Capel-Sanchez, M., ... Fierro, J. L. G. (2018). Cermets Ni/(Ce0.9Ln0.1O1.95) (Ln = Gd, La, Nd and Sm) prepared by solution combustion method as catalysts for hydrogen production by partial oxidation of methane. *Int. J. Hydrogen Energ.*, *43*, 16834–16845.

Alvarez, J., Valderrama, G., Pietri, E., Perez-Zurita, M. J., Urbina de Navarro, C., Sousa-Aguiar, E. F., & Goldwasser, M. R. (2011). Ni–Nb-based mixed oxides precursors for the dry reforming of methane. *Top. Catal.*, *54*, 170–178.

Amjad, U. S., Moncada-Quintero, C. W., Ercolino, G., Italiano, C., Vita, A., & Specchia, S. (2019). Methane steam reforming on Pt/CeO2 catalyst: Effect of daily start-up and shut-down on long term stability of catalyst. *Ind. Eng. Chem. Res.*, *58*, 16395–16406.

Arkatova, L. A., Kasatsky, N. G., Maximov, Y. M., Pakhnutov, O. V., & Shmakov, A. N. (2018). Intermetallides as the catalysts for carbon dioxide reforming of methane. *Catal. Today*, *299*, 303–316.

Bartholomew, C. H., & Farrauto, R. J. (2006). *Fundamentals of Industrial Catalytic Processes* (2nd ed.). New Jersey: Wiley-Interscience.

Bera, P. (2019). Solution combustion synthesis as a novel route to preparation of catalysts. *Int. J. Self-Propag. High-Temp Synth.*, *28*, 77–109.

Carlos, E., Martnis, R., Fortunato, E., & Branquinho, R. (2020). Solution combustion synthesis: Towards a sustainable approach for metal oxides. *Chem. Eur. J.* https://doi.org/10.1002/chem.20200067.

Cesario, M. R., Barros, B. S., Courson, C., Melo, D. M. A., & Kiennemann, A. (2015). Catalytic performances of Ni–CaO–mayenite in CO2 sorption enhanced steam methane reforming. *Fuel Process. Technol.*, *131*, 247–253.

Chorkendorff, I., & Niemantsverdriet, J. W. (2003). *Concepts of Modern Catalysis and Kinetics*. Weinheim: Wiley-VCH.

Choudhary, T. V., Aksoylu, E., & Goodman, D. W. (2003). Nonoxidative activation of methane. *Catal. Rev.*, *45*, 151–203.

Cipiti, F., Pino, L., Vita, A., Lagana, M., & Recupero, V. (2013). Experimental investigation on a methane fuel processor for polymer electrolyte fuel cells. *Int. J. Hydrogen Energ.*, *38*, 2387–2397.

Cross, A., Miller, J. T., Danghyan, V., Mukasyan, A. S., & Wolf, E. E. (2019). Highly active and stable Ni–Cu supported catalysts prepared by combustion synthesis for hydrogen production from ethanol. *Appl. Catal. A: Gen.*, *572*, 124–133.

de Jong, K. P. (2009). General aspects. In K. P. de Jong (Ed.), *Synthesis of Solid Catalysts* (pp. 3–10). Weinheim: Wiley-VCH.

Delmon, B. (2007). Catalysts for new uses: Needed preparation advances. In J. Regalbuto (Ed.), *Catalyst Preparation. Science and Engineering* (pp. 449–463). London: CRC Press.

Dincer, I., & Acar, C. (2015). Review and evaluation of hydrogen production methods for better sustainability. *Int. J. Hydrog. Energy*, *40*, 11094–11111.

Edwards, P. P., Kuznetsov, V. L., David, W. I. F., & Brandon, N. P. (2008). Hydrogen and fuel cells: Towards a sustainable energy future. *Energy Policy, 36*, 4356–4362.

Enger, B. C., Lødeng, R., & Holmen, A. (2008). A review of catalytic partial oxidation of methane to synthesis gas with emphasis on reaction mechanisms over transition metal catalysts. *Appl. Catal. A: Gen., 346*, 1–27.

Erri, P., Nader, J., & Varma, A. (2008). Controlling combustion wave propagation for transition metal/alloy/cermet foam synthesis. *Adv. Mater., 20*, 1243–1245.

Erria, P., Dinka, P., & Varma, A. (2006). Novel perovskite-based catalysts for autothermal JP-8 fuel reforming. *Chem. Eng. Sci., 61*, 5328–5333.

Fang, X., Zhang, X., Guo, Y., Chen, M., Liu, W., Xu, X., ... Li, C. (2016). Highly active and stable Ni/Y2Zr2O7 catalysts for methane steam reforming: On the nature and effective preparation method of the pyrochlore support. *Int. J. Hydrogen Energ., 41*, 11141–11153.

Galletti, C., Djinovic, P., Specchia, S., Batista, J., Levec, J., Pintar, A., & Specchia, V. (2011). Influence of the preparation method on the performance of Rh catalysts on CeO2 for WGS reaction. *Catal. Today, 176*, 336–339.

Garcia, V., Caldes, M. T., Joubert, O., Sierra-Gallego, G., Batiot-Dupeyrat, C., Piffard, Y., & Moreno, J. A. (2008). Dry reforming of methane over nickel catalysts supported on the cuspidine-like phase Nd4Ga2O9. *Catal. Today, 133–135*, 231–238.

González-Cortés, S., & Imbert, F. E. (2018). *Advanced Solid Catalysts for Renewable Energy Production*. IGI Global.

González-Cortés, S. L., Rugmini, S., Xiao, T., Green, M. L. H., Rodulfo-Baechler, S. M., & Imbert, F. E. (2014). Deep hydrotreating of different feedstocks over a highly active Al2O3-supported NiMoW sulfide catalyst. *Appl. Catal. A: Gen., 475*, 270–281.

González-Cortés, S. L., Xiao, T.-C., & Green, M. L. H. (2006). Urea–matrix combustion method: A versatile tool for the preparation of HDS catalysts. *Studies Surf. Sci. Catal., 162*, 817–824.

González-Cortés, S. L., Xiao, T., Costa, P. M. F. J., Fontal, B., & Green, M. L. H. (2004). Urea–organic matrix method: An alternative approach to prepare Co-MoS2/γ-Al2O3 HDS catalyst. *Appl. Catal. A: Gen., 270*, 209–222.

González-Cortés, S. L., Xiao, T., Lin, T.-W., & Green, M. L. H. (2006). Influence of double promotion on HDS catalysts prepared by urea-matrix combustion synthesis. *Appl. Catal. A: Gen., 302*, 264–273.

González-Cortés, S. L., Xiao, T., Rodulfo-Baechler, S. M. A., & Green, M. L. H. (2005). Impact of the urea–matrix combustion method on the HDS performance of Ni–MoS2/γ–Al2O3 catalysts. *J. Molec. Catal. A: Chem., 240*, 214–225.

González-Cortés, S., Rodulfo-Baechler, S. R., & Imbert, F. E. (2014b). Solution combustion method: A convenient approach for preparing Ni promoted Mo and MoW sulphide hydrotreating catalysts. In J. M. Grier (Ed.), *Combustion: Types of Reactions, Fundamental Processes and Advanced Technologies* (pp. 269–288): Nova Science Publishers, Inc.

González-Cortés, S., Slocombe, D., Xiao, T., Aldawsari, A., Yao, B., Kuznetsov, V. L., ... Edwards, P. P. (2016). Wax: A benign hydrogen-storage material that rapidly releases H2-rich gases through microwave-assisted catalytic decomposition. *Scientific Reports, 6*(35315).

He, F., Chen, J., Liu, S., Huang, Z., Wei, G., Wang, G., ... Zhao, K. (2019). La1-xSrxFeO3 perovskite-type oxides for chemical looping steam methane reforming: Identification of the surface elements and redox cyclic performance. *Int. J. Hydrogen Energ., 44*, 10265–10276.

Hennings, U., & Reimert, R. (2007). Investigation of the structure and the redox behavior of gadolinium doped ceria to select a suitable composition for use as catalyst support in the steam reforming of natural gas. *Appl. Catal. A: Gen., 325*, 41–49.

Holladay, J. D., Hu, J., King, D. L., & Wang, Y. (2009). An overview of hydrogen production technologies. *Catal. Today, 139*, 244–260.

Jie, X., González-Cortés, S., Xiao, T., Wang, J., Yao, B., Slocombe, D. R., ... Edwards, P. P. (2017). Rapid production of high-purity hydrogen fuel through microwave-promoted deep catalytic dehydrogenation of liquid alkanes with abundant metals. *Angew. Chem. Int. Edit., 56*(34), 10170–10173.

Jie, X., González-Cortés, S., Xiao, T., Yao, B., Wang, J., Slocombe, R. D., ... Edwards, P. P. (2019). The decarbonisation of petroleum and other fossil hydrocarbon fuels for the facile production and safe storage of hydrogen. *Energy Environ. Sci., 12*, 238–249.

Kalamaras, M., & Efstathiou, A. (2013). Hydrogen production technologies: Current state and future developments. *Conference Papers in Energy*, 1–9.

Kapteijn, F., Martin, G. B., & Moulijn, J. A. (1993). Catalytic reaction engineering. In J. A. Moulijn, P. W. N. M. van Leeuwen, & R. A. van Santen (Eds.), *Catalysis: An Integrated Approach to Homogeneous, Heterogeneous and Industrial Catalysis*. Amsterdam: Elsevier.

Khani, Y., Bahadoran, F., Soltanali, S., & Ahari, J. S. (2018). Hydrogen production by methanol steam reforming on a cordierite monolith reactor coated with Cu–Ni/LaZnAlO4 and Cu–Ni/γ-Al2O3 catalysts. *Res. Chem. Intermed., 44*, 925–942.

Khort, A., Podboloto, K., Serrano-García, R., & Gun'ko, Y. K. (2017). One-step solution combustion synthesis of pure Ni nanopowders with enhanced coercivity: The fuel effect. *J. Solid State Chem., 253*, 270–276.

Kingsley, J. J., & Patil, K. C. (1988). A novel combustion process for the synthesis of fine particle α-alumina and related oxide materials. *Mater. Lett., 6*, 427–432.

Kothari, R., Buddhi, D., & Sawhney, R. L. (2008). Comparison of environmental and economic aspects of various hydrogen production methods. *Renew. Sust. Energ. Rev., 12*, 553–563.

Kumar, A., Wolf, E. E., & Mukasyan, A. S. (2011). Solution combustion synthesis of metal nanopowders: Nickel-reaction pathways. *AIChE J., 57*, 2207–2214.

Kuvshinov, D. G., Kurmashov, P. B., Bannov, A. G., Popov, M. V., & Kuvshinov, G. G. (2019). Synthesis of Ni-based catalysts by hexamethylenetetramine-nitrates solution combustion method for co-production of hydrogen and nanofibrous carbon from methane. *Int. J. Hydrogen Energ.,* *44*, 16271–16286.

Lamba, N., Gupta, R., Modak, J. M., & Madras, G. (2019). ZnO catalyzed transesterification of Madhuca indica oil in supercritical methanol. *Fuel, 242,* 323–333.

Lima, M.-W., Yong, S.-T., & Chai, S.-P. (2014). Combustion-synthesized nickel-based catalysts for the production of hydrogen from steam reforming of methane. *Energy Procedia, 61,* 910–913.

Luiz, A., Marinho, A., Rabelo-Neto, R. C., Noronha, F. B., & Mattos, L. V. (2016). Steam reforming of ethanol over Ni-based catalysts obtained from LaNiO3 and LaNiO3/CeSiO2 perovskite-type oxides for the production of hydrogen. *Appl. Catal. A: Gen., 520,* 53–64.

Maksimov, Y. M., Kirdyashkin, A. I., & Arkatova, L. A. (2013). Conversion of methane on catalysts obtained via self-propagating high-temperature synthesis. *Catal. Ind., 5,* 245–252.

Mandizadeh, S., Sadri, M., & Salavati-Niasari, M. (2017). Sol-gel auto combustion synthesis of BaFe18O27 nanostructures for adsorptive desulfurization of liquid fuels. *Int. J. Hydrogen Energ., 42,* 12320–12326.

Munnik, P., de Jongh, P. E., & de Jong, K. P. (2015). Recent developments in the synthesis of supported catalysts. *Chem. Rev., 115,* 6687–6718.

Muradov, N., & Veziroğlu, N. (2008). Green path from fossil-based to hydrogen economy: An overview of carbon-neutral technologies. *Int. J. Hydrogen Energ., 33,* 6804–6839.

Muradov, N. Z., & Veziroglu, T. N. (2005). From hydrocarbon to hydrogen–carbon to hydrogen economy. *Int. J. Hydrogen Energ., 30,* 225 –237.

Navlani-García, M., Mori, K., Kuwahara, Y., & Yamashita, H. (2018). Recent strategies targeting efficient hydrogen production from chemical hydrogen storage materials over carbon-supported catalysts. *NPG Asia Mater., 10,* 277–292.

Nayebzadeh, H., Saghatoleslami, N., Haghighi, M., & Tabasizadeh, M. (2019). Catalytic activity of KOH–CaO–Al2O3 nanocomposites in biodiesel production: Impact of preparation method. *Int. J. Self-Propag. High Temp. Synth., 28,* 18–27.

Nguyen, T.-S., Postole, G., Loridant, S., Bosselet, F., Burel, L., Aouine, M., ... Piccolo, L. (2014). Ultrastable iridium–ceria nanopowders synthesized in one step by solution combustion for catalytic hydrogen production. *J. Mater. Chem. A, 2,* 19822–19832.

Ojeda-Nino, O. H., Gracia, F., & Daza, C. (2019). Role of Pr on Ni–Mg–Al mixed oxides synthesized by microwave-assisted self-combustion for dry reforming of methane. *Ind. Eng. Chem. Res., 58,* 7909–7921.

Patil, K. C., Hedge, M. S., Rattan, R., & Aruna, S. T. (2008). *Chemistry of Nanocrystalline Oxide Materials: Combustion Synthesis, Properties and Applications.* Singapore: World Scientific.

Pino, L., Italiano, C., Vita, A., Lagana, M., & Recupero, V. (2017). Ce0.70La0.20Ni0.10O2-δ catalyst for methane dry reforming: Influence of reduction temperature on the catalytic activity and stability. *Appl. Catal. B, 218*, 779–792.

Pino, L., Vita, A., Cipiti, F., Lagana, M., & Recupero, V. (2006). Performance of Pt/CeO2 catalyst for propane oxidative steam reforming. *Appl. Catal. A: Gen., 306*, 68–77.

Pino, L., Vita, A., Cipiti, F., Lagana, M., & Recupero, V. (2011). Hydrogen production by methane tri-reforming process over Ni–ceria catalysts: Effect of La-doping. *Appl. Catal. B, 104*, 64–73.

Pino, L., Vita, A., Cordaro, M., Recupero, V., & Hegde, M. S. (2003). A comparative study of Pt/CeO2 catalysts for catalytic partial oxidation of methane to syngas for application in fuel cell electric vehicles. *Appl. Catal. A: Gen., 243*, 135–146.

Podbolotov, K. B., Khort, A. A., Tarasov, A. B., Trusov, G. V., Roslyakov, S. I., & Mukasyan, A. S. (2017). Solution combustion synthesis of copper nanopowders: The fuel effect. *Combust. Sci. Technol., 189*, 1878–1890.

Postole, G., Nguyen, T.-S., Aouine, M., Gelin, P., Cardenas, L., & Piccolo, L. (2015). Efficient hydrogen production from methane over iridium-doped ceria catalysts synthesized by solution combustion. *Appl. Catal. B, 166–167*, 580–591.

Potemkin, D. I., Aghayan, M., Uskov, S. I., Snytnikov, P. V., Kamboj, N., Rodríguez, M. A., ... Sobyanin, V. A. (2018). Fibrous alumina-based Ni-CeO2 catalyst: Synthesis, structure and properties in propane pre-reforming. *Mater. Lett., 215*, 35–37.

Pudukudy, M., Yaakob, Z., Jia, Q., & Takriffa, M. S. (2018). Catalytic decomposition of undiluted methane into hydrogen and carbon nanotubes over Pt promoted Ni/CeO2 catalysts. *New J. Chem., 42*, 14843–14856.

Regalbuto, J. (2007). *Catalyst Preparation Science and Engeering*. London: CRC Press.

Rodriguez-Sulbaran, P. J., Lugo, C. A., Perez, M. A., González-Cortés, S. L., D'Angelo, R., Rondon, J., ... Del Castillo, H. L. (2018). Dry reforming of methane on LaSrNiAl perovskite-type structures synthesized by solution combustion. In S. González-Cortés & F. E. Imbert (Eds.), *Advanced Solid Catalysts for Renewable Energy Production* (pp. 242–266). Hershey: IGI Global.

Sajjadia, S. M., & Haghighi, M. (2019). Combustion vs. hybrid sol-gel-plasma surface design of coke-resistant Copromoted Ni-spinel nanocatalyst used in combined reforming of CH4/CO2/O2 for hydrogen production. *Chem. Eng. J., 362*, 767–782.

Scharz, J. A., Contescu, C., & Contescu, A. (1995). Methods for preparation of catalytic materials. *Chem. Rev., 95*, 477–510.

Schulz, C., Roy, S. C., Wittich, K., d'Alnoncourt, R. N., Linke, S., Strempel, V. E., ... Rosowski, F. (2019). αII-(V1-xWx)OPO4 catalysts for the selective oxidation of n-butane to maleic anhydride. *Catal. Today, 333*, 113–119.

Shikina, N. V., Yashnik, S. A., Gavrilova, A. A., Ishchenko, A. V., Dovlitova, L. S., Khairulin, S. R., & Ismagilov, Z. R. (2019). Effect of glycine addition on physicochemical and catalytic properties of Mn, Mn–La and Mn–Ce monolithic catalysts prepared by solution combustion synthesis. *Catal. Lett.,* *149*, 2535–2551.

Specchia, S., Ercolino, G., Karimi, S., Italiano, C., & Vita, A. (2017). Solution combustion synthesis for preparation of structured catalysts: A mini-review on process intensification for energy applications and pollution control. *Int. J. Self-Propag. High Temp. Synth., 26*, 166–186.

Specchia, S., Finocchio, E., Busca, G., & Specchia, V. (2010). Combustion synthesis. In M. Lackner, F. Winter, & A. K. Agarwal (Eds.), *Handbook of Combustion* (pp. 439–472). Weinheim: Wiley-VCH Verlag GmbH & Co. KGaA.

Tajuddin, M. M., Ideris, A., & Ismail, M. (2019). In situ glycine-nitrate combustion synthesis of Ni-La/SiO2 catalyst for methane cCracking. *Ind. Eng. Chem. Res., 58*, 531–538.

Tappan, B. C., Huynh, M. H., Hiskey, M. A., Chavez, D. E., Luther, E. P., Mang, J. T., & Son, S. F. (2006). Ultralow-Density nanostructured metal foams: Combustion synthesis, morphology, and composition. *J. Am. Chem. Soc., 128*, 6589–6594.

Toscani, L. M., Zimicz, M. G., Martins, T. S., Lamas, D. G., & Larrondox, S. A. (2018). In situ X-ray absorption spectroscopy study of CuO–NiO/CeO2–ZrO2 oxides: Redox characterization and its effect in catalytic performance for partial oxidation of methane. *RSC Adv., 8*, 12190–12203.

Trusov, G. V., Tarasov, A. B., Goodilin, E. A., Rogachev, A. S., Roslyakov, S. I., Rouvimov, S., ... Mukasyan, A. S. (2016). Spray solution combustion synthesis of metallic hollow microspheres. *J. Phys. Chem. C, 120*, 7165–7171.

Tsodikov, M. V., Teplyakov, V. V., Fedotov, A. S., Kozitsyna, N. Y., Bychkov, V. Y., Korchak, V. N., & Moiseev, I. I. (2011). Dry reforming of methane on porous membrane catalytic systems. *Russ. Chem. Bull. Int. Ed., 60*, 55–62.

Uvarov, V. I., & Borovinskaya, I. P. (2013). SHS-produced catalytically active porous membranes containing nickel nanoparticles. *Int. J. Self-Propag. High Temp. Synth., 22*, 232–233.

Vita, A., Italiano, C., Ashraf, M. A., Pino, L., & Specchia, S. (2018). Syngas production by steam and oxy-steam reforming of biogas on monolith-supported CeO2-based catalysts. *Int. J. Hydrogen Energ., 43*, 11731–11744.

Vita, A., Italiano, C., Fabiano, C., Pino, L., Lagana, M., & Recupero, V. (2016). Hydrogen-rich gas production by steam reforming of n-dodecane. Part I: Catalytic activity of Pt/CeO2 catalysts in optimized bed configuration. *Appl. Catal. B, 199*, 350–360.

Vita, A., Italiano, C., Pino, L., Frontera, P., Ferraro, M., & Antonucci, V. (2018). Activity and stability of powder and monolith-coated Ni/GDC catalysts for CO2 methanation. *Appl. Catal. B, 226*, 384–395.

Vita, A., Italiano, C., Pino, L., Lagana, M., & Recupero, V. (2017). Hydrogen-rich gas production by steam reforming of n-dodecane. Part II: Stability, regenerability and sulfur poisoning of low loading Rh-based catalyst. *Appl. Catal. B-Environ.*, *218*, 317–326.

Vita, A., Pino, L., Cipiti, F., Lagana, M., & Recupero, V. (2010). Structured reactors as alternative to pellets catalyst for propane oxidative steam reforming. *Int. J. Hydrogen Energ.*, *35*, 9810–9817.

Voskanyan, A. A., & Chan, K. Y. (2018). Scalable synthesis of three-dimensional meso/macroporous NiO with uniform ultralarge randomly packed mesopores and high catalytic activity for soot oxidation. *ACS Appl. Nano Mater.*, *1*, 556–563.

Wang, C., Yu, F., Zhu, M., Tang, C., Dong, L., & Dai, B. (2018). Synthesis of both powdered and preformed MnOx–CeO2–Al2O3 catalysts by self-propagating high-temperature synthesis for the selective catalytic reduction of NOx with NH3. *ACS Omega*, *3*, 5692–5703.

Wang, H., Dong, X., Zhao, T., Yu, H., & Li, M. (2019). Reforming of methane over bimetallic Ni-Co catalyst prepared from La(CoxNi1-x)0.5Fe0.5O3 perovskite precursor: Catalytic activity and coking resistance. *Appl. Catal. B*, *245*, 302–313.

Wang, K., Huang, J., Li, W., Huang, J., Sun, D., Ke, X., & Li, Q. (2019). Role of mineral nutrients in plant-mediated synthesis of three-dimensional porous LaCoO3. *Ind. Eng. Chem. Res.*, *58*, 8555–8564.

Wang, M., Zhao, T., Li, M., & Wang, H. (2017). Perovskite La2(NiCu)O4 catalyst precursors for dry reforming of methane: Effects of Cu-substitution on carbon resistance. *RSC Adv.*, *7*, 41847–41854.

WEO. (2018). International Energy Agency. Retrieved from https://www.iea.org/weo2018/.

Xanthopoulou, G., Marinou, A., Karanasios, K., & Vekinis, G. (2017). Combustion synthesis during flame spraying ("CAFSY") for the production of catalysts on substrates *Coatings*, *7*, 14.

York, A. P. E., Xiao, T., & Green, M. L. H. (2003). Brief overview of the partial oxidation of methane to synthesis gas. *Top. Catal.*, *22*, 345–358.

Yousefi, S., Haghighi, M., & Vahid, B. R. (2019). Role of glycine/nitrates ratio on structural and texture evolution of MgO based nanocatalyst fabricated by hybrid microwave-impregnation method for biofuel production. *Energy Convers. Manag.*, *182*, 251–261.

Index